Time

Perception of Temporal Patterns by the Recursive Model

By

Behzad Ghorbani

© Behzad Ghorbani, 2024.

All rights reserved.

No part of this book may be reproduced, distributed, or transmitted in any form or by any means, including photocopying, recording, or other electronic or mechanical methods, without the prior written permission of the publisher, except in the case of brief quotations embodied in critical reviews and certain other non-commercial uses permitted by copyright law.

This work is a product of hybrid collaboration, authored by Behzad Ghorbani and refined through the Hyper Hybrid Intellect model, with AI acting as the joint hybrid editor. This prototype combines human creativity and thought leadership with advanced analytical enhancement and structural optimisation (including Comparative Matrix Analysis, Mirroring Hemisphere Fractal Analysis, Totalisation of Details, and Recursive Realism Perspective), ensuring a uniquely crafted and rigorously developed narrative.

First Edition

Published by Amazon.com

10 9 8 7 6 5 4 3 2 1

November 2024, London

Paperback: ISBN: 9798345127094
Hardcover: ISBN: 9798345145159

CONTENTS

    Introduction   6

1. The Fractal Brain: Structural Foundations of the Recursive Mind   9

2. The Recursive Nature of Human Cognition: Time and Thought Loops   12

3. Time Perception and Memory: A Recursive Dance of Past and Present   17

4. Time in the Physical Universe: Relativity, Thermodynamics, and the Recursive Mind   23

5. Circadian Rhythms, Biological Time, and the Recursive Mind   29

6. Consciousness, Awareness, and Time: The Present Moment in the Recursive Mind   35

7. Time as a Cultural Construct: The Social and Philosophical Dimensions of Time   42

8. Technology, Artificial Intelligence, and the Acceleration of Time Perception   49

9. The Biological Foundations of Time Perception: Neural and Cellular Mechanisms   56

10. Philosophical Theories of Time: Metaphysics, Perception, and the Recursive Mind   63

11	Time and Memory: The Reconstruction of Temporal Narratives	70
12	Emotion and Time Perception, The Recursive Interplay Between Feelings and Temporal Experience	77
13	Cultural and Societal Constructs of Time: The Social Frameworks of Temporal Experience	87
14	Cultural and Societal Constructs of Time: The Social Frameworks of Temporal Experience	94
15	Attention, Consciousness, and Cognitive Control: Time Perception and Complex Decision-Making	101
16	Creativity, Imagination, and Narrative Thinking: Time Perception and the Construction of Temporal Reality	108
17	The Philosophy of Time: Metaphysical Perspectives and Recursive Processing	115
18	Time and Memory: Constructing and Reconstructing Temporal Narratives	124
19	Emotion and Time: The Recursive Loops of Emotional Memory and Time Perception	130
20	Cultural Frameworks and Time: The Influence of Society on Temporal Perception	136
21	Social Interaction and Time: The Collective Experience of Temporal Awareness	143
22	Technology, Artificial Intelligence, and the Digital Age: Transforming Our Experience of Time	150
23	The Future of Time: Emerging Technologies and Time-Bending Possibilities	156

24	Time and Consciousness: Synthesising Insights from the Recursive Mind Model	162
25	Practical Applications of the Recursive Mind Model: Education, Therapy, and Technology	168
26	Future Research and Development: Expanding the Boundaries of the Recursive Mind Model	175
27	The Recursive Mind and the Nature of Reality: Implications for Human Experience and the Universe	182
28	Reflections and Future Directions: The Ongoing Journey of Understanding Time and the Recursive Mind	188
	Conclusion	192
	References	197

Introduction: The Recursive Mind and the Nature of Time

The Recursive Mind is a model that seeks to illuminate one of the most enigmatic and fascinating aspects of human cognition: how we perceive, understand, and interact with the concept of time. Time is not merely an external measurement; it is interwoven into the very fabric of human experience, from our subjective consciousness of time passing, to our ability to predict the future, to the scientific frameworks that explore time's role as a fundamental dimension of the physical universe.

This model, the Recursive Mind, provides a fresh approach to explaining how the brain handles complex cognitive tasks, including time. At its core, this model suggests that the brain is equipped with a powerful recursive capacity, the ability to loop back, process patterns, and refine information. This recursive processing is not merely a method of thinking but a structural feature embedded in the very architecture of the brain. Through these recursive loops, the brain generates insights, refines its understanding, and allows for the synthesis of concepts across multiple dimensions, including time.

The focus of this book is to examine time from every conceivable angle, drawing from philosophy, science, psychology, biology, and cultural studies, and cross-examine these perspectives using the Recursive Mind model. In doing so, we seek to determine whether the brain's recursive nature can truly explain the vast and diverse ways in which humans understand and experience time. The Recursive Mind model not only offers a new understanding of neural processing but also provides a framework for addressing fundamental questions about time itself, such as:

- How does the recursive nature of the brain influence our perception of the past, present, and future?
- Can the recursive processes in the brain explain why time feels relative, why it seems to speed up or slow down depending on our emotional and psychological state?
- How does the brain's ability to process patterns recursively help us handle abstract concepts like time dilation, entropy, or even the thermodynamic arrow of time?
- What role does memory, consciousness, and the neural architecture of the brain play in shaping our experience of time?
- Can the recursive, fractal-based structure of the cerebral cortex explain multiscale time perception, bridging the gap between short-term perception and long-term temporal understanding?

The Recursive Mind model offers the potential not only to address these questions but also to generate new hypotheses and theories about time. By delving deeply into how recursion operates within the brain, this book will explore how this cognitive process might lead to new discoveries and insights in the realms of time perception, neuroscience, and even cosmology. The journey we are about to embark on will take us through philosophical conundrums, scientific puzzles, and psychological mysteries, all viewed through the lens of neural recursion.

Chapter 1: The Fractal Brain: Structural Foundations of the Recursive Mind

The human brain is an extraordinary organ, not merely because of its computational power, but because of the unique and intricate ways it is structured. One of the most fascinating aspects of the brain's architecture is its fractal nature, a geometric feature where a structure exhibits self-similarity at various scales. This feature is most visible in the cerebral cortex, the folded outer layer of the brain responsible for many of our higher cognitive functions. This fractal geometry is not merely an aesthetic feature or a consequence of evolutionary development; rather, it is key to the brain's ability to process information in recursive loops, which is central to the Recursive Mind model.

To understand how the fractal structure of the brain supports recursive thinking and time perception, we must first explore what fractals are. In essence, a fractal is a structure that repeats itself across different scales. If one were to zoom into a small section of a fractal, one would find that it looks remarkably similar to the larger structure. This characteristic of self-similarity across scales is not merely a mathematical curiosity but is critical to how the brain processes information. In a fractal structure, each part contains information about the whole, allowing the brain to store, process, and refine information across different levels of complexity.

In the case of the cerebral cortex, this folding pattern allows the brain to maximise its surface area, providing more space for neuronal connections without increasing the volume of the brain. This structural efficiency is not just about compactness; it allows for multi-scale processing, where the brain can handle both local and global information at the same time. In the context of time perception, this means that the brain is capable of simultaneously processing immediate sensory input (such as the perception of a present moment) and long-term temporal patterns (such as the concept of a lifetime or the sequence of historical events).

The Recursive Mind model posits that this fractal structure is not just a physical feature but is also key to how the brain organises and processes time. Time itself is often understood as having a fractal-like structure, where short moments of time resemble longer periods in their fundamental nature. For example, we can think of a single second, a minute, an hour, or a year as self-similar units of time that differ in scale but share a fundamental structure. The brain, with its recursive fractal architecture, is able to process these different scales of time simultaneously, giving us the ability to perceive both short-term events and long-term durations in a coherent way.

One of the key aspects of the Recursive Mind is its ability to engage in recursive processing. Recursion refers to the process by which the brain takes its output and loops it back as input for further refinement. This recursive processing is essential for handling complex tasks, especially in relation to time perception. For instance, when we experience the present moment, the brain is constantly refining its perception by referencing past experiences and future expectations. The recursive nature of this process allows us to bind together sensory information into a coherent whole, creating a seamless experience of the present moment.

Furthermore, the brain's ability to process time recursively is not limited to the present. It also allows us to reflect on the past and project into the future. Memory plays a crucial role in this process, as it provides the raw material for the brain's recursive loops. Each time we recall a memory, the brain mirrors it and integrates it into the current context, creating a recursive feedback loop between past experiences and present cognition. Similarly, the brain's ability to simulate the future, to anticipate events and make predictions, is also based on recursive processing, where the brain uses past patterns to project forward in time.

The fractal nature of the brain's architecture is critical to its ability to handle these multiple scales of time. Just as fractals repeat their patterns across different levels of magnification, the brain's recursive processes allow it to navigate time at multiple levels. We can understand this as the brain's ability to zoom in and zoom out on different temporal scales, moving fluidly from the immediate present to broader temporal frameworks, such as decades or even entire lifetimes. This dynamic flexibility is one of the hallmarks of the Recursive Mind and is key to understanding how we process time both subjectively and objectively.

The fractal architecture of the brain and its recursive nature provide the foundation for a new way of understanding how we perceive time. In the chapters that follow, we will explore how this model applies across different realms, including psychology, neuroscience, and physics, and we will develop new hypotheses about the role of recursion in shaping our understanding of time.

Chapter 2: The Recursive Nature of Human Cognition: Time and Thought Loops

At the heart of the Recursive Mind model lies the understanding that human cognition operates through the continuous looping and refining of information. This process of recursion allows the brain to handle not only the sensory world but also abstract concepts, introspection, and time. The brain's ability to engage in self-referential thought, reflecting on its own processes, plays a crucial role in how we perceive and navigate the concept of time. In this chapter, we will explore the cognitive mechanisms that underpin recursive thinking and examine how these processes give rise to our understanding of past, present, and future.

Recursion in cognition can be likened to the concept of feedback loops, where the output of one mental process is fed back as input for further processing. This continuous loop is what allows humans to reflect on the past, engage in prospection (the ability to anticipate and simulate the future), and refine their present understanding. Unlike simple linear processing, recursion enables a kind of dynamic flexibility that is crucial for handling complex temporal constructs, such as the passage of time, memory, and the anticipation of future events.

One of the most significant features of recursive cognition is its ability to bridge time. The brain does not experience time in isolation; instead, it is constantly referencing past experiences and future possibilities as it processes the present moment. This recursive integration of past, present, and future is what allows us to have a coherent sense of continuity, a self that exists through time. Without this recursive framework, our experience of time would be fragmented, and we would lack the ability to situate ourselves in a broader temporal context.

To understand how the brain achieves this, we must first explore the concept of temporal binding. Temporal binding refers to the brain's ability to synchronize sensory inputs that arrive at slightly different times, creating a unified perception of the present. For example, when we hear a sound, see a movement, and feel a touch, these sensory inputs do not reach the brain simultaneously. However, the brain is able to bind them together into a single, coherent experience of the present moment. This process of binding is inherently recursive, the brain must continuously loop back and refine its understanding of the present by integrating multiple inputs over time.

Temporal binding is particularly important in the context of time perception. Time is not something we perceive directly; rather, it is a construct that the brain generates by synthesizing sensory information. The Recursive Mind model suggests that this process is recursive because the brain is constantly updating its perception of time based on new inputs and past experiences. For example, when we are in a state of flow, deeply immersed in an activity, time seems to speed up because the brain is recursively compressing sensory inputs into a more streamlined experience. Conversely, when we are bored or anxious, time seems to slow down because the brain is recursively expanding its processing loops, giving more attention to the passage of each moment.

One of the most profound aspects of recursive cognition is its ability to engage in mental time travel. Unlike other species, humans have the unique ability to mentally project themselves into the future and reflect on the past. This ability to simulate future scenarios and recall past experiences is a direct result of the brain's recursive loops. When we imagine the future, the brain is not simply constructing a random sequence of events; rather, it is recursively referencing patterns from past experiences to generate predictions about what might happen next. This process is recursive because each prediction is refined by feeding back the brain's output as input for further simulation, creating a continuous loop of anticipation and expectation.

Similarly, when we recall a past event, the brain is engaging in a recursive reconstruction of that memory. Each time we retrieve a memory, the brain loops back to the original event but also integrates new information, emotions, and context into that memory. This means that memory is not static; it is constantly being updated and refined through recursive processing. This explains why our memories of past events often change over time, each time we recall the memory, we are adding new layers of context and interpretation, creating a recursive loop that continuously shapes our understanding of the past.

One of the most intriguing aspects of recursive cognition is its role in the sense of self. The self is not a fixed entity; rather, it is a dynamic construct that is continually shaped and reshaped by recursive processes in the brain. Our sense of self is deeply tied to our perception of time because we experience ourselves as existing through time, from our memories of past experiences to our hopes and plans for the future. The Recursive Mind model suggests that this temporal self is constructed through recursive loops that continuously integrate past, present, and future experiences into a coherent narrative.

The brain's ability to construct a temporal narrative is key to understanding how we make sense of time. Humans are natural storytellers, and one of the most important stories we tell is the story of our own lives. This autobiographical narrative is recursively constructed by the brain, which weaves together memories, immediate experiences, and future goals into a continuous thread. This narrative gives us a sense of continuity, the feeling that we are the same person today as we were in the past and will be in the future. Without recursive processing, this narrative would be fragmented, and our sense of self would be disjointed.

In the context of time perception, recursive cognition allows us to zoom in and out on different temporal scales. For example, we can reflect on the immediate past (what happened a few minutes ago) or the distant past (what happened decades ago), and we can anticipate the near future (what will happen later today) or the far future (what might happen years from now). The Recursive Mind model suggests that this ability to navigate different scales of time is made possible by the brain's fractal structure, which allows for multi-scale processing. Just as a fractal exhibits self-similarity across different scales, the brain's recursive loops allow it to process time at both micro (momentary) and macro (lifelong) levels.

The fractal nature of time perception is evident in the way we experience cycles and repetition. For example, the brain is able to process the daily cycle of waking and sleeping (circadian rhythms), as well as longer cycles such as the seasons or the stages of life. These temporal cycles are processed recursively by the brain, which recognises patterns and loops back to integrate them into its ongoing perception of time. This recursive processing allows us to make sense of repeating events and to anticipate the future based on past patterns.

The Recursive Mind model provides a powerful framework for understanding how the brain handles temporal complexity. By engaging in recursive loops, the brain is able to integrate past experiences, present perceptions, and future predictions into a coherent understanding of time. This recursive processing is essential not only for our perception of time but also for our ability to navigate time, to plan, reflect, and make decisions that are informed by both past and future considerations.

As we move forward in this book, we will continue to explore how recursive cognition shapes our understanding of time across different domains, from memory to consciousness to scientific theories. By examining the recursive nature of human cognition, we can gain new insights into the brain's remarkable ability to process time and generate meaning across temporal dimensions.

Chapter 3: Time Perception and Memory: A Recursive Dance of Past and Present

The relationship between time perception and memory is perhaps one of the most compelling aspects of the human experience. Our ability to recall past events, project into the future, and remain grounded in the present moment all rely on the brain's extraordinary capacity to weave together threads of temporal experience. The Recursive Mind model offers a unique lens through which to understand this intricate process, highlighting how our perceptions of time are the result of recursive feedback loops between past memories, current sensory input, and future simulations.

This chapter explores the neuroscientific and cognitive foundations of time perception, focusing on how memory systems, especially episodic memory, create a recursively informed sense of time. By understanding how the brain's recursive processes integrate past and present, we can gain deeper insight into why time sometimes feels fast, slow, or even disjointed.

The Brain's Temporal Fabric: How Memory and Perception Interact

Time is not something we perceive in the same way we perceive objects or sounds. Rather, it is a constructed experience, created by the brain as it integrates sensory information, memory, and cognitive expectations. At the core of this integration lies a recursive process by which the brain takes in present stimuli, references stored memories, and creates a temporal framework that allows us to understand the flow of time.

Memory plays a central role in this process. When we experience something new, the brain encodes that event and stores it in episodic memory, which is responsible for recalling specific events from our personal past. However, memories are not static entities stored like files in a computer. Every time we recall a memory, the brain reconstructs it, and this reconstruction is subject to recursive feedback. Each new experience that resembles a previous one triggers the brain to reference earlier memories, update them with new information, and refine its understanding of both the present and the past.

This recursive interplay between memory and current perception is what allows us to have a sense of temporal continuity. For example, when we walk into a room we've been in before, the brain references past memories of the room while simultaneously processing the current sensory information. This creates a blended experience of time, where the past and present are experienced as part of a unified whole. Without this recursive integration of memory and perception, our experience of time would feel fragmented, and we would be unable to maintain a consistent narrative of our lives.

One of the most fascinating implications of the Recursive Mind model is that it suggests time perception is not a passive process but an active, dynamic construction. The brain is constantly working to synchronize past experiences with the present moment, creating a coherent temporal framework that allows us to navigate time with ease. This is why, when our memory systems are disrupted, such as in cases of amnesia or dementia, our perception of time becomes disoriented, fragmented, or even non-existent.

The Past in the Present: How Memory Shapes Time Perception

The brain's ability to time travel, to move back and forth between past memories and present experiences, is a hallmark of its recursive nature. Every moment of present experience is informed by a vast network of past experiences, memories, and contextual information. This allows us to perceive time as a flow rather than a series of disconnected moments.

One way the brain achieves this is through a process known as temporal binding, where it combines sensory information from different moments to create a sense of temporal continuity. Temporal binding is a recursive process because the brain is constantly looping back to reference prior moments, integrating them into the current perception of time. This is why we often experience a sense of smoothness in time, where moments seem to blend into one another rather than appearing as discrete chunks.

Memory plays a crucial role in temporal binding. Each time we recall a memory, the brain loops back to that past moment and integrates it with the present. This allows us to compare the present moment to past experiences, drawing on patterns and similarities that help us navigate the flow of time. For example, when we meet a friend for the first time in years, our brain instantly references past memories of that friend, helping us to contextualize the present moment and create a seamless temporal experience. This recursive loop between past and present is what allows us to maintain a continuous sense of self across time.

However, this process is not without its quirks. Because memory is reconstructed each time it is recalled, it is subject to modifications and distortions. Each time we retrieve a memory, we are not accessing an exact replica of the original event but rather a recreation of that event, filtered through our current cognitive state, emotional context, and new experiences. This explains why memories often change over time, each time they are recalled, the brain updates them with new information, creating a recursive loop that refines the memory while also altering it.

This process of memory modification has profound implications for how we perceive time. Since our memories are constantly being reshaped, our sense of temporal continuity is also in flux. Events that seemed distant in the past may feel more recent when they are frequently recalled, while other events may fade into the background of our consciousness. The brain's recursive modification of memory is what allows us to maintain a fluid and flexible understanding of time, but it also means that our perception of time is subjective and malleable.

Time Dilation and Compression: How Emotions and Cognition Influence Time Perception

Another fascinating aspect of the brain's recursive processing is its ability to distort time based on emotional and cognitive states. Time is not always experienced at a constant pace; instead, it can speed up or slow down depending on our level of attention, focus, and emotional engagement. This phenomenon, known as time dilation and time compression, can be explained through the Recursive Mind model as the result of recursive loops that either compress or expand our perception of time.

When we are deeply immersed in an activity, such as during a state of flow, the brain's recursive loops become more efficient, allowing us to process sensory information more quickly and with fewer distractions. As a result, time seems to speed up because the brain is compressing its processing loops, focusing only on the most essential information and filtering out extraneous details. This creates the sensation that time is passing rapidly, as the brain is not spending as much cognitive effort on tracking each moment.

Conversely, when we are bored or anxious, the brain's recursive loops become less efficient, leading to a sense of time dilation. In these states, the brain's processing loops become longer and more drawn out, as the brain focuses more on internal thoughts and external distractions. This creates the sensation that time is dragging on, as the brain is spending more cognitive effort on tracking each moment. The recursive nature of this process means that the brain is constantly reassessing its perception of time, adjusting it based on the current cognitive and emotional context.

This phenomenon is particularly evident in situations of high emotional intensity, where time seems to stretch or shrink depending on the emotional weight of the moment. For example, during moments of extreme fear or excitement, time often seems to slow down as the brain's recursive loops focus intensely on the details of the experience, processing each moment in greater depth. On the other hand, when we are experiencing joy or pleasure, time often seems to fly by, as the brain's recursive loops streamline the processing of sensory information, allowing us to focus on the positive emotions rather than the passage of time itself.

Hypothesis: Can Recursive Processes Explain Temporal Illusions?

The Recursive Mind model offers a compelling hypothesis for why humans experience temporal illusions, distortions in the perception of time. Temporal illusions, such as the sensation that time is speeding up or slowing down, can be understood as the result of recursive modifications in the brain's processing loops. When the brain compresses or expands its recursive loops, it alters the way we perceive the duration of events, leading to time distortions.

One testable hypothesis is that neural efficiency, the speed and accuracy with which the brain processes recursive loops, directly correlates with the experience of time dilation and time compression. By studying brain activity during states of flow, boredom, fear, and pleasure, we can examine how the brain's recursive processes are modulated in different emotional and cognitive states. If the Recursive Mind model is correct, we would expect to find that faster recursive loops (more efficient neural processing) correlate with time compression, while slower recursive loops (less efficient processing) correlate with time dilation.

Moreover, the model suggests that disruptions in the brain's recursive loops, such as those caused by neurological disorders, may lead to more extreme forms of temporal illusions. For example, individuals with schizophrenia or depersonalization disorders often report experiencing time as fragmented or disjointed, which may be explained by disruptions in the brain's ability to engage in recursive processing. Future research could explore whether therapies aimed at restoring or enhancing recursive processing in the brain could help alleviate these temporal distortions.

In conclusion, the Recursive Mind model provides a powerful framework for understanding the complex relationship between memory, emotion, and time perception. The brain's ability to loop back to past experiences, modify them, and integrate them with present sensory input is key to creating a coherent sense of time. At the same time, the brain's recursive processes allow for flexibility in time perception, explaining why time can feel fast, slow, or distorted depending on our cognitive and emotional state.

As we move forward, we will continue to explore how the brain's recursive nature shapes our understanding of time across other domains, including consciousness, scientific theories of time, and biological rhythms. By examining these different aspects of time through the lens of recursion, we can gain new insights into one of the most fundamental aspects of human experience.

Chapter 4: Time in the Physical Universe: Relativity, Thermodynamics, and the Recursive Mind

To fully grasp how humans understand time, it is crucial to cross-examine the brain's recursive cognitive processes against the backdrop of time as a physical phenomenon. This chapter explores time in the context of physics, focusing on Einstein's theory of relativity, quantum mechanics, and the thermodynamic arrow of time. Through these lenses, we will investigate whether the Recursive Mind model can explain how the brain processes these abstract scientific concepts, integrating them into a coherent framework that can be mapped onto our subjective experience of time.

Time in the physical universe is often described in terms that challenge our everyday perceptions. In Newtonian mechanics, time was thought to be an absolute, continuous flow, ticking away uniformly across the universe. However, the emergence of Einstein's theory of relativity fundamentally changed our understanding of time, showing that it is relative to an observer's frame of reference, influenced by speed and gravity. Time can dilate or contract, depending on how fast we are moving or how close we are to a massive object. Simultaneously, thermodynamics introduced the concept of an arrow of time, a one-way direction toward entropy and disorder, which seems to contrast with the reversibility of physical laws on a smaller scale.

Relativity and Time Perception: The Brain's Ability to Process Non-Linear Time

Einstein's theory of relativity introduced the idea that time is not a fixed and absolute entity but is intimately connected to space, forming a four-dimensional continuum known as spacetime. The most counterintuitive aspect of relativity is the fact that time is relative, meaning that it can flow at different rates depending on an object's speed or gravitational field. For instance, as an object approaches the speed of light, time dilates, slowing down relative to an observer at rest. This means that for someone travelling near the speed of light, hours may pass while centuries pass for someone stationary.

The question that arises is: How does the brain, a biological entity rooted in everyday experience, come to understand and grapple with these counterintuitive scientific realities? The Recursive Mind model offers a potential explanation by suggesting that the brain's ability to process time recursively allows it to construct an abstract understanding of time's relativity, even though we do not directly perceive time dilation in our daily lives.

The brain processes temporal sequences by continually looping back and integrating information from both past experiences and present input. This recursive processing allows the brain to handle non-linearities in time, such as those introduced by relativity, by extrapolating from known patterns and constructing abstract models that can accommodate these seemingly paradoxical ideas. In essence, the brain uses recursive simulation to create a mental model of time that is flexible enough to accommodate the scientific understanding of time as a relative construct.

Consider how we intuitively grasp the concept of time dilation: while we do not experience it directly, our brains can simulate it through recursive loops that mirror the underlying logic of relativity. By referencing and refining its understanding of temporal flow, the brain can generate models that allow us to comprehend how time might behave under different physical conditions, such as near the event horizon of a black hole. The brain's recursive nature allows it to integrate these abstract scientific principles into its broader understanding of time, even though such phenomena are far removed from our everyday experience.

Thermodynamics and the Arrow of Time: Reconciling Entropy with Cognitive Processing

In the realm of thermodynamics, time takes on a different meaning. The second law of thermodynamics states that entropy, or disorder, in a closed system will always increase over time. This gives rise to the concept of the arrow of time, the idea that time moves in one direction, from order to disorder, and that this direction is irreversible. This stands in stark contrast to many physical laws at the quantum level, which are time-symmetric, meaning they work the same whether time moves forward or backward.

The concept of entropy and the arrow of time poses unique challenges for how we understand and perceive time. The brain, which processes time as a continuous, forward-moving flow, must reconcile this unidirectional nature of time with the fact that many of the physical processes it deals with (such as neural activity, motor control, and perception) operate in reversible ways. How, then, does the brain manage to process the idea of irreversible time, the sense that time only moves forward and that past events cannot be undone?

The Recursive Mind model suggests that the brain can handle this paradox through hierarchical recursion. At a micro-level, the brain may process time in a reversible way, through processes such as sensory perception or motor control, where actions can be quickly corrected or adjusted. However, at a macro-level, the brain processes time as moving irrevocably forward, integrating the concept of entropy into its broader temporal framework.

For example, our sense of time's irreversibility is strongly influenced by memory and experience. Once a memory is encoded, it becomes a part of the brain's temporal structure, and while the details of that memory can be modified or recalled differently, the fact of the memory's existence remains fixed. This creates a cognitive sense of time moving forward, where past events are preserved in memory and cannot be undone, much like the entropy in the universe.

Moreover, the brain's ability to process time recursively may help explain why we intuitively understand the arrow of time in relation to causality. Recursive processes allow the brain to simulate the consequences of events, projecting future outcomes based on past actions. This ability to project forward in time, while referencing and refining past experiences, creates a temporal narrative that aligns with the irreversible nature of thermodynamic time.

Quantum Mechanics and Time: Is the Brain Equipped to Handle Quantum Temporal Paradoxes?

In the realm of quantum mechanics, time becomes even more perplexing. At the quantum level, particles do not follow the linear flow of time that we experience in our macroscopic world. Quantum events can occur in ways that seem to defy the traditional concept of time, such as in the phenomenon of quantum entanglement, where two particles appear to communicate instantaneously across vast distances, seemingly bypassing the constraints of time and space.

The brain, which evolved to navigate the macroscopic world, is not naturally equipped to process the quantum mechanical view of time. However, the Recursive Mind model offers a potential framework for understanding how the brain might simulate and model these abstract concepts. Through recursive processing, the brain can generate hypothetical models that allow it to handle the paradoxes inherent in quantum mechanics, such as superposition and non-locality.

One way in which the brain achieves this is through its ability to hold multiple possibilities in mind at once, a form of cognitive superposition. When we think about quantum phenomena, the brain does not collapse these possibilities into a single outcome but instead recursively loops through different scenarios, holding competing ideas in tension. This recursive flexibility allows the brain to simulate quantum temporal paradoxes, even though they defy the traditional flow of time as we experience it.

Moreover, the brain's recursive nature allows it to integrate quantum temporal concepts with its broader understanding of time. While quantum mechanics challenges the brain's intuitive grasp of time, recursion enables the brain to abstract these ideas and construct models that are consistent with both classical and quantum views of time. This ability to navigate multiple frameworks and reconcile macro-level temporal flow with quantum indeterminacy is a testament to the brain's recursive capacity.

The Recursive Mind's Capacity for Abstract Temporal Understanding

One of the greatest strengths of the Recursive Mind model is its ability to abstract temporal concepts across multiple domains. Time in physics behaves very differently from time in our subjective experience, yet the brain is able to construct models that reconcile these disparate views. Through recursive loops, the brain can simulate both relativistic and thermodynamic time, while also generating models that accommodate quantum temporal paradoxes.

This recursive capacity suggests that the brain is not bound by its biological origins, it has evolved to be a flexible, abstract-thinking machine capable of navigating the most complex ideas in time and space. The ability to engage in recursive simulation allows the brain to handle time dilation, entropy, and quantum paradoxes, integrating them into a coherent framework that can be applied to both scientific reasoning and everyday cognition.

Hypothesis: Can Recursive Simulation Help Bridge the Gap Between Subjective and Objective Time?

A key hypothesis that emerges from the Recursive Mind model is that the brain's recursive loops allow it to bridge the gap between subjective time (the time we experience in our daily lives) and objective time (the time described by physics). By recursively referencing past experiences, present input, and future projections, the brain can generate models that integrate the non-linearities and paradoxes of time described in both relativity and quantum mechanics.

This hypothesis could be tested by examining how individuals understand abstract temporal concepts under different cognitive conditions. For instance, we could investigate how people process time dilation when presented with scientific explanations versus everyday time distortions, such as time dilation during emotional states. By tracking brain activity, we could identify the recursive loops involved in abstract temporal reasoning and explore how they relate to both subjective and scientific models of time.

The Brain as a Temporal Bridge

In conclusion, the Recursive Mind model provides a robust framework for understanding how the brain navigates the complexities of time in the physical universe. Whether dealing with the relativity of time, the thermodynamic arrow, or the paradoxes of quantum mechanics, the brain's recursive processing allows it to simulate and abstract these concepts, bridging the gap between subjective experience and scientific theory.

As we continue our exploration, we will further investigate how the brain's recursive capacity shapes our understanding of time in different realms, including biological rhythms and consciousness. By examining time through the lens of recursion, we can gain deeper insights into how the brain creates meaning from the mystery of time itself.

Chapter 5: Circadian Rhythms, Biological Time, and the Recursive Mind

While the perception of time can be an abstract, cognitive construct, the human body operates under a more biologically grounded sense of time. This biological time is governed by the circadian rhythms, the internal processes that regulate the sleep-wake cycle, metabolism, hormonal secretion, and other vital physiological functions. These rhythms are synchronised with environmental cues, such as the cycle of day and night, and they are essential for maintaining overall health and wellbeing.

However, these biological clocks are not merely passive mechanisms; they engage in a form of recursive processing that allows the body and brain to synchronise with the external environment and adapt to changes. In this chapter, we explore how biological time intersects with the Recursive Mind model, examining how recursive processes in the brain interact with circadian rhythms, creating a dynamic system that influences both physiology and cognition. Additionally, we will examine how disruptions in biological rhythms, whether through shift work, jet lag, or artificial light exposure, affect time perception, mental health, and overall cognitive function.

The Brain's Biological Clock: How Circadian Rhythms Influence Time Perception

At the heart of biological time is the circadian system, which consists of a network of molecular clocks that regulate the timing of cellular processes across the body. The primary pacemaker of this system is located in the suprachiasmatic nucleus (SCN) of the hypothalamus, a small region of the brain that responds to light and darkness cues. The SCN synchronises with the light-dark cycle of the external world and communicates this timing information to other parts of the brain and body, ensuring that physiological processes occur at the appropriate times of day.

The recursive nature of the brain's interaction with the circadian system becomes evident when we consider how the SCN continuously adapts to environmental changes. For example, when exposed to artificial light at night or when travelling across time zones, the SCN receives conflicting signals from the environment. In response, the brain engages in a recursive loop of adjusting its internal clocks, gradually realigning them with the new light-dark cycle.

This process of realignment is not instantaneous; it occurs through a recursive feedback loop where the brain must continuously refine its timekeeping mechanisms based on incoming sensory data. As the SCN recalibrates, the brain synchronises physiological functions, such as sleep patterns, metabolism, and cognitive alertness, with the new environmental cues. This recursive adjustment illustrates how the brain's temporal cognition is not only subjective but also intricately linked to biological rhythms that govern our daily lives.

Memory, Sleep, and the Role of Recursion in Biological Time

One of the most significant ways in which circadian rhythms influence time perception is through their effect on memory and cognitive function, particularly in relation to sleep. Sleep is not merely a state of rest; it is a vital process during which the brain engages in memory consolidation, organising and refining information learned during the day. This process is inherently recursive, as the brain repeatedly loops through memories and sensory data, selectively storing important information while discarding irrelevant details.

Sleep occurs in stages, alternating between rapid eye movement (REM) sleep and non-REM sleep, and these stages follow a circadian rhythm. During REM sleep, the brain engages in heightened neuronal activity, which is associated with vivid dreams and the processing of emotional memories. Non-REM sleep, on the other hand, is crucial for deep rest and the consolidation of procedural memory, skills and tasks learned during the day. The cycling between these stages follows a recursive pattern, as the brain loops back and forth between REM and non-REM sleep throughout the night.

This recursive sleep cycle is essential for maintaining temporal continuity in memory. Studies have shown that when sleep is disrupted, whether by irregular sleep patterns, insomnia, or sleep disorders, the brain's ability to consolidate memories is impaired, leading to memory lapses, difficulty concentrating, and a distorted perception of time. This demonstrates the critical role of recursive sleep cycles in maintaining our ability to perceive and navigate time effectively.

Moreover, the internal clock that governs sleep is sensitive to changes in external time cues, such as light exposure, but it also relies on internal feedback loops to maintain synchrony with environmental rhythms. The brain's ability to track time during sleep and wakefulness, adjusting to changes in light and darkness, is a prime example of how biological time is managed through recursive processes.

Biological Time and the Aging Process: The Role of Circadian Disruption

As we age, the precision of the body's circadian system begins to decline, leading to disruptions in sleep patterns, metabolic processes, and even cognitive function. This decline in circadian regulation can have profound effects on how we perceive time, particularly in relation to the acceleration of subjective time that many people experience as they grow older.

The Recursive Mind model suggests that part of the brain's ability to maintain a coherent perception of time relies on the synchronisation of biological rhythms with external environmental cues. As these rhythms become disrupted with age, the brain's recursive processing loops may also lose efficiency, leading to a distorted sense of time. Older adults often report that time seems to pass more quickly, a phenomenon that may be related to the decline of recursive temporal processes as the brain's ability to synchronise with biological clocks diminishes.

Moreover, circadian disruption is not limited to the aging process. Individuals who work night shifts, travel across time zones frequently, or are exposed to irregular light-dark cycles (as in modern societies dominated by artificial light) often experience chronic circadian misalignment. This misalignment can lead to sleep disorders, cognitive impairments, and even increased risk for chronic diseases, such as diabetes, heart disease, and depression. The brain's recursive ability to adjust to changes in time is challenged under these conditions, leading to long-term dysregulation of both biological time and time perception.

The Recursive Integration of Biological and Cognitive Time

While biological time is governed by the circadian system, it cannot be fully separated from cognitive time, our conscious perception of the passage of moments, hours, and days. The Recursive Mind model proposes that the brain operates as a synchronisation system, integrating both internal biological rhythms and external environmental cues to create a unified sense of time.

This recursive integration is key to understanding how the brain maintains temporal coherence. For example, during the day, the brain must balance cognitive demands (such as attention, memory, and decision-making) with biological needs (such as hunger, fatigue, and alertness). The SCN and other circadian pacemakers communicate with regions of the brain responsible for cognitive function, ensuring that our mental states are aligned with the body's physiological rhythms. This communication occurs through recursive feedback loops, allowing the brain to adapt to changing environmental conditions while maintaining internal synchrony.

In situations where biological rhythms are disrupted, such as during periods of sleep deprivation or jet lag, the brain's recursive integration of biological and cognitive time becomes misaligned, leading to a sense of disorientation and temporal distortion. Under these conditions, individuals often experience difficulties with attention, short-term memory, and temporal tracking, all of which are directly influenced by the recursive interaction between circadian rhythms and cognitive processing.

Hypothesis: Can the Recursive Mind Explain Circadian Synchronisation Across Different Time Scales?

A potential hypothesis that arises from the Recursive Mind model is that the brain's recursive loops enable it to synchronise time perception across different biological and cognitive time scales. For instance, the ability to perceive time at the level of seconds (e.g., reacting to a moving object) may be linked to the same recursive processes that govern daily rhythms (e.g., the sleep-wake cycle) or even seasonal changes (e.g., mood fluctuations in response to daylight variation).

This hypothesis could be tested by examining how disruptions in circadian rhythms, such as those caused by shift work, jet lag, or seasonal affective disorder (SAD), impact both short-term and long-term time perception. By analysing brain activity during periods of circadian misalignment, researchers could explore how recursive processes in the brain are affected by biological time disruptions and whether these disruptions alter the brain's ability to track and synchronise time across different temporal scales.

Additionally, future research could investigate whether enhancing circadian regulation, through exposure to natural light, optimised sleep schedules, or circadian-aligned therapies, could improve time perception, cognitive function, and overall mental health, particularly in individuals who experience chronic circadian disruption.

Conclusion: The Interaction of Biological and Cognitive Time in the Recursive Mind

In conclusion, the Recursive Mind model offers a compelling framework for understanding how the brain integrates biological rhythms with cognitive time, creating a coherent sense of temporal flow. Through recursive processes, the brain synchronises its internal clocks with external cues, allowing it to maintain a dynamic balance between physiological needs and cognitive demands.

The recursive interaction between the brain's circadian system and its cognitive processing allows for adaptive flexibility, enabling individuals to navigate both the biological cycles that regulate sleep, wakefulness, and metabolism, as well as the cognitive processes that govern memory, attention, and time perception. This interaction is key to maintaining both mental health and temporal coherence, particularly in environments where biological rhythms are constantly challenged by modern lifestyles.

As we proceed, we will continue to explore how the brain's recursive nature influences other realms of time, particularly in relation to consciousness and self-awareness, as well as the broader cultural and philosophical implications of time as a human construct.

Chapter 6: Consciousness, Awareness, and Time: The Present Moment in the Recursive Mind

One of the most intriguing aspects of time is our conscious awareness of the present moment. While time itself is often seen as an abstract and objective phenomenon, our subjective experience of time is deeply tied to our consciousness. Consciousness enables us to perceive the passage of time, to situate ourselves within the flow of events, and to engage in reflective thought about both the past and the future. But what exactly is happening in the brain when we experience the present moment, and how does the Recursive Mind model explain this phenomenon?

In this chapter, we will explore the intersection of consciousness and time perception, focusing on how the brain's recursive loops contribute to the experience of now, the elusive moment between the past and future. We will also examine how attention, awareness, and self-reflection are dynamically interconnected in the recursive processes that allow us to engage with time in a conscious and meaningful way. Moreover, we will explore how altered states of consciousness, such as those induced by meditation, drugs, or extreme emotion, can distort the perception of time, shedding light on the malleable nature of temporal awareness.

The Elusive Present: What is "Now" in the Context of the Recursive Mind?

The present moment, or now, is often regarded as the only point in time where we can truly be aware and conscious. The past is stored in memory, and the future exists only in anticipation, but now is the point at which we experience immediate reality. Yet, defining or even pinpointing what constitutes the present moment is a challenge. Neurological studies suggest that our perception of now is not a static point in time but is instead a moving window of time that stretches across milliseconds or even seconds, creating a sense of temporal coherence.

From the perspective of the Recursive Mind model, the brain's experience of the present moment is the result of a continuous process of recursion, in which the brain integrates sensory inputs with internal states, such as memory and expectation, to create a coherent, fluid perception of time. This recursive process is essential for giving us a sense of continuity and temporal unity. Without it, we would experience time as fragmented, with no clear connection between moments.

At its core, the experience of the present moment involves a binding process, where the brain takes multiple streams of sensory information, each arriving at slightly different times, and integrates them into a single, unified perception. For instance, visual, auditory, and tactile signals may all reach the brain at different speeds, yet we do not perceive a disjointed reality. The Recursive Mind ensures that these signals are synchronised through recursive loops, allowing us to experience the present moment as a cohesive whole.

Moreover, internal states such as emotional responses, bodily sensations, and cognitive expectations are continuously looped back into the brain's processing of the present. This recursive feedback allows the brain to adjust its perception of the present based on both external stimuli and internal experiences. For instance, when we are in a highly emotional state, such as fear or excitement, the recursive loops in the brain may expand or contract the perceived duration of the present moment, leading to temporal distortions such as time seeming to slow down or speed up.

Attention and Temporal Binding: How Focus Shapes Our Perception of Time

Attention plays a central role in how we perceive the present moment, and it is through recursive loops that the brain directs and sustains attention over time. When we focus on a particular task or event, the brain's neural circuits loop back on themselves, continuously refining and adjusting the focus of attention based on sensory feedback and cognitive goals. This recursive process of attention regulation is key to how we experience time because it determines which aspects of the present moment are brought into awareness and how those aspects are integrated into our broader sense of temporal flow.

For example, when we are deeply engrossed in an activity, such as reading a book or playing a game, our attention becomes highly focused, and the recursive loops in the brain work to compress the passage of time, filtering out extraneous information and focusing only on the task at hand. This explains why time seems to fly by during periods of intense focus: the brain is recursively narrowing its attention to a single activity, creating a streamlined experience of the present.

Conversely, when we are distracted or unable to focus, such as during moments of boredom or anxiety, the brain's recursive loops expand, allowing more sensory information and internal thoughts to flood into consciousness. This can create the sensation that time is dragging on, as the brain struggles to maintain a coherent focus and instead loops through multiple competing streams of information. The Recursive Mind model suggests that these variations in attention are not simply cognitive states but are directly linked to our experience of time. As attention contracts or expands, so too does our perception of the present moment.

The phenomenon of temporal binding, where the brain integrates sensory inputs over time, also depends on attention. In order to experience the present moment as coherent, the brain must recursively loop back to integrate previous moments with the current sensory data. This creates a smeared present, where our experience of now stretches across several milliseconds or even seconds. The more focused our attention, the more compressed this window becomes, and the more fluid and immediate our experience of the present.

Consciousness and Time Distortion: Altered States of Awareness and Temporal Perception

One of the most compelling aspects of consciousness is its ability to alter our perception of time. In states of normal waking consciousness, time often seems to flow at a relatively steady pace, punctuated by moments of heightened attention or distraction. However, in altered states of consciousness, such as those induced by meditation, drugs, extreme emotional states, or even traumatic experiences, time can become distorted in ways that reveal the brain's flexible and recursive processing of temporal information.

For instance, during meditative states, practitioners often report a sense of timelessness or the expansion of the present moment, where time seems to slow down or even stand still. In these states, the brain's recursive loops may decelerate, allowing the individual to focus intensely on the present without the usual distraction of past memories or future concerns. By recursively looping back to the immediate sensory input, the brain can create a prolonged sense of now, where the past and future recede into the background of awareness.

On the other hand, in states of extreme fear or trauma, individuals often report that time seems to stretch, with moments feeling drawn out or as though they are moving in slow motion. This phenomenon, known as time dilation, occurs because the brain's recursive loops are working overtime to process every detail of the immediate environment. In a threatening situation, the brain enhances its recursive processing to ensure that all sensory input is analysed in greater detail, allowing the individual to respond to the threat more effectively. The result is a perception of time as slowed down, with each moment feeling longer than usual.

Psychoactive substances, such as hallucinogens, can also profoundly affect time perception. Individuals under the influence of such substances often report that time seems to lose structure or that they are experiencing multiple timelines simultaneously. The Recursive Mind model offers a potential explanation for these experiences by suggesting that hallucinogens disrupt the brain's normal recursive loops, leading to a disintegration of the usual temporal structure. Without the recursive processes that bind sensory input and memory into a coherent stream, time may feel disjointed, with moments no longer following one another in a linear fashion.

The Self and the Temporal Continuum: How Recursion Constructs the Sense of Time

Our experience of time is deeply intertwined with our sense of self. We perceive ourselves as existing through time, moving from past to present to future, and this sense of temporal continuity is essential for maintaining a coherent self-concept. The Recursive Mind model suggests that this temporal self is constructed through continuous loops that integrate memories of the past, sensory input from the present, and anticipations of the future.

These recursive loops allow us to maintain a narrative of the self, where our past experiences are continuously updated and integrated into our current identity. For example, when we reflect on a significant life event, the brain recursively loops back to the memory of that event, recontextualising it based on our current understanding and experiences. This recursive process is not simply a matter of recalling the past but of reconstructing it, updating it with new information and perspectives.

Similarly, our ability to project into the future is governed by recursive loops that take past patterns and current goals and use them to simulate potential future scenarios. This recursive simulation allows us to engage in planning, decision-making, and self-reflection, all of which are essential for navigating time. By integrating past, present, and future into a continuous loop, the brain constructs a temporal self that can move fluidly across time.

Hypothesis: Can the Recursive Mind Explain Variations in Temporal Awareness?

A key hypothesis that emerges from the Recursive Mind model is that variations in temporal awareness, such as the distortions experienced in altered states of consciousness, are the result of modulations in the brain's recursive loops. When the brain's recursive loops are accelerated (as in heightened states of attention), time seems to compress, creating a sense of temporal flow. Conversely, when the recursive loops decelerate (as in meditation or trauma), time seems to expand, leading to a stretched or elongated perception of the present.

This hypothesis could be tested by examining brain activity during states of focused attention, distraction, and altered consciousness, using neuroimaging techniques to observe the speed and complexity of recursive loops. By correlating these findings with subjective reports of time perception, researchers could explore how changes in recursive processing affect the experience of time.

The Recursive Mind and the Fluidity of Time in Consciousness

In conclusion, the Recursive Mind model provides a robust explanation for how the brain constructs the present moment and navigates the flow of time. Through recursive loops, the brain integrates sensory input, memory, and attention, creating a dynamic and flexible sense of temporal awareness. This recursive processing is key to understanding not only how we perceive time in normal waking consciousness but also how time becomes distorted in altered states and extreme experiences.

As we move forward, we will continue to explore how the brain's recursive nature influences other dimensions of time, such as the role of cultural constructs and philosophical theories in shaping our understanding of time's nature and meaning.

Chapter 7: Time as a Cultural Construct: The Social and Philosophical Dimensions of Time

While time is often considered a universal concept, its meaning and significance can vary dramatically across different cultures, philosophical traditions, and societies. How people perceive time, organise their daily lives around it, and even define the past, present, and future can depend largely on the cultural context in which they are embedded. Time is not just a scientific abstraction or a biological rhythm; it is also a cultural construct, shaped by societal values, historical developments, and philosophical perspectives.

In this chapter, we will explore how cultural constructs of time interact with the Recursive Mind model, particularly in relation to how individuals internalise and navigate these constructs. We will examine Western and Eastern philosophies of time, the linear versus cyclical perception of time, and how technological advancements and modern lifestyles have transformed our understanding and experience of time. By looking at time through a cultural lens, we can gain insights into how the brain's recursive processing adapts to these societal frameworks, helping individuals align their cognitive processes with broader cultural narratives.

Linear and Cyclical Time: Western and Eastern Cultural Perspectives

One of the most prominent distinctions in the cultural understanding of time is the difference between linear and cyclical models. In Western societies, time is predominantly viewed as linear, a straight line that extends from the past, moves through the present, and reaches into the future. This linear model is deeply embedded in Western thought, from the Christian narrative of creation and salvation to the scientific and historical understanding of progress and development. Time is seen as directional, moving forward inexorably, and individuals are expected to organise their lives around this progression, striving toward future goals and achievements.

In contrast, many Eastern cultures adopt a cyclical view of time. In Buddhism, Hinduism, and Taoism, time is seen as a series of repeating cycles, governed by the forces of birth, death, and rebirth. The seasons of nature, the cycles of the moon, and the daily rhythms of life all reflect this belief that time does not move in a straight line but instead revolves in an endless loop. In this worldview, the focus is often less on the future as a linear destination and more on achieving balance and harmony within the ongoing cycles of existence.

The Recursive Mind model provides a framework for understanding how these different cultural conceptions of time are internalised by the brain. In societies where time is viewed as linear, individuals are likely to develop a cognitive model of time that emphasises forward movement, with the brain recursively organising experiences into a progressive narrative. This forward-looking mentality is reflected in the brain's planning mechanisms, where recursive loops enable individuals to anticipate future events and work toward long-term goals. The brain's recursive nature supports the linear progression of thought and action, continuously projecting into the future based on past experiences.

In contrast, individuals living in cyclical cultures may develop a recursive model of time that emphasises repetition and continuity rather than progression. The brain's recursive loops would focus more on pattern recognition, enabling individuals to recognise cyclical patterns in nature, life, and society. This focus on recurrence rather than linear advancement may shape the brain's temporal processing, aligning it more closely with the cyclical rhythms of the natural world.

Time and Social Organisation: From Agricultural Societies to the Digital Age

The way societies organise their activities around time has profound implications for how individuals perceive and internalise time. In pre-industrial agricultural societies, time was largely governed by the natural cycles of the seasons, the rising and setting of the sun, and the rhythms of planting and harvesting. Individuals' daily routines were tied to the earth's natural cycles, and time was perceived more organically, as something that flowed according to nature's rhythms.

The Industrial Revolution dramatically changed this relationship with time. With the invention of mechanical clocks, the regulation of the workday became rigid and precise, with hours and minutes governing labour, production, and daily activities. Time became a commodity, something that could be measured, saved, and spent. The brain had to adapt to this new way of organising life, shifting from a natural, cyclical relationship with time to a highly structured, linear one. In many ways, the recursive processes in the brain were restructured to accommodate the increased emphasis on punctuality, efficiency, and the constant movement forward in time.

In the digital age, this relationship with time has evolved even further. With the advent of the internet, smartphones, and global communication, time has become compressed and accelerated. People now experience time as moving at a faster pace, with constant streams of information, notifications, and demands for attention. The brain's recursive loops are increasingly tasked with managing multiple timelines, work, personal life, social media, and global news, often simultaneously. This can lead to a sense of time compression, where days, weeks, or even months seem to blur together.

The Recursive Mind model suggests that the brain's ability to process multiple streams of information is rooted in its recursive architecture. However, as the demands of modern life increase, these recursive loops are stretched to their limits, potentially leading to cognitive overload and temporal distortion. Individuals may feel as though time is slipping away more quickly because the brain's recursive processes are overwhelmed by the sheer volume of information and activity that must be processed and integrated.

The Social Construction of Time: Calendars, Clocks, and Temporal Norms

One of the most powerful tools that societies use to structure time is the calendar. The development of calendars allowed societies to organise activities around days, weeks, months, and years, creating a shared sense of time that could be applied to everything from religious ceremonies to agricultural cycles to business transactions. These calendars were not just practical tools; they also reflected deeper cultural beliefs about the nature of time and the universe.

For example, the Gregorian calendar, used widely in the Western world, reflects a linear view of time, with a fixed beginning (the birth of Christ) and a continuous progression into the future. In contrast, the Mayan calendar was cyclical, based on complex astronomical observations and organised around repeating cycles of time, such as the Long Count, which spanned thousands of years.

The Recursive Mind model can help explain how these societal constructs of time are internalised by individuals. When a society adopts a specific calendar or system of timekeeping, it imposes a temporal structure on its members, shaping how they think about the past, present, and future. The brain's recursive processes adapt to this structure, learning to synchronise its internal clocks with the socially constructed calendar. For example, the recursive loops responsible for planning and memory will align with the cultural understanding of annual cycles, such as holidays, seasons, and important life events.

Similarly, the invention of mechanical clocks and the subsequent development of time zones standardised time across vast regions, enabling global synchronisation of activities. This had a profound impact on the brain's recursive processes, particularly in how individuals managed time zones and adjusted their internal clocks to align with global time standards. The recursive nature of the brain's processing allowed individuals to adapt to these temporal norms, creating a shared, synchronised experience of time across different geographic locations.

Philosophical Conceptions of Time: Subjective Experience and Objective Reality

In addition to cultural constructs, philosophy has long wrestled with the nature of time, often questioning whether time is a subjective experience or an objective reality. Immanuel Kant famously argued that time (and space) is not something that exists independently of human perception but is instead a framework that the mind imposes on sensory experiences. According to Kant, time is a product of human cognition, a necessary condition for organising our experiences into a coherent sequence.

The Recursive Mind model aligns with Kant's philosophy in suggesting that time is largely a mental construct, created through the brain's recursive processes. These processes allow the brain to structure sensory input into past, present, and future, creating the illusion of temporal continuity. Without the brain's recursive loops to bind together moments of sensory experience, we would not be able to perceive time as a continuous flow. This idea resonates with phenomenological approaches to time, which emphasise the subjective experience of time as something deeply intertwined with consciousness.

In contrast, Newtonian and Einsteinian physics treat time as an objective reality, a dimension of the universe that exists independently of human perception. While Newton viewed time as an absolute entity that ticks away uniformly regardless of external conditions, Einstein showed that time is relative to the observer's frame of reference, influenced by velocity and gravity. The brain's ability to reconcile these scientific models of time with its subjective experience of time is one of the most remarkable aspects of human cognition. Through recursive loops, the brain can abstract these scientific concepts, integrating them with the lived experience of time, even though they may challenge our intuitive understanding of time's flow.

Hypothesis: Can the Recursive Mind Explain Cultural Differences in Time Perception?

A key hypothesis arising from the Recursive Mind model is that cultural differences in time perception, such as the distinction between linear and cyclical views of time, are the result of the brain's ability to adapt its recursive processes to align with cultural norms. When individuals grow up in a culture that emphasises linear time, their brains develop recursive loops that reinforce goal-oriented, forward-thinking behaviour. Conversely, in cultures that emphasise cyclical time, the brain's recursive processes may become more attuned to pattern recognition and repetition.

This hypothesis could be tested by comparing brain activity in individuals from different cultural backgrounds while they engage in tasks related to time perception, planning, and memory. By examining how recursive loops function in these tasks, researchers could explore whether there are cultural differences in how the brain processes time, and whether these differences correspond to the cultural constructs of time that individuals internalise.

The Social Construction of Time and the Recursive Mind

In conclusion, time is not just a physical or biological phenomenon; it is also a cultural construct that shapes how individuals experience and navigate the world. The Recursive Mind model provides a powerful framework for understanding how the brain's recursive processes adapt to these cultural constructs, allowing individuals to align their cognitive processes with broader societal narratives of time.

As we move forward, we will explore how the brain's recursive nature continues to influence our understanding of time, particularly in the context of technological advancements, artificial intelligence, and the future of time perception in a rapidly changing world.

Chapter 8: Technology, Artificial Intelligence, and the Acceleration of Time Perception

The rapid development of technology in the 21st century has profoundly changed how humans experience and interact with time. With the advent of artificial intelligence, instant communication, and global connectivity, time has become increasingly compressed, and our perception of its flow has shifted. Individuals now live in an age where information is available at their fingertips, where responses are expected in milliseconds, and where the pace of life is constantly accelerating. In this chapter, we will examine how technological advancements have influenced time perception, focusing particularly on how artificial intelligence (AI) and digital environments interact with the Recursive Mind model.

By exploring how the brain adapts to the ever-increasing demands for speed and efficiency in the digital age, we will investigate whether these technological advancements are reshaping the brain's recursive loops and altering the way individuals experience and process time. Additionally, we will examine the potential for AI to replicate, enhance, or even alter human time perception, raising important questions about the future of cognitive interaction with artificial systems.

The Acceleration of Time in the Digital Age: From Biological Time to Digital Time

The rise of the digital age has led to a phenomenon often described as the acceleration of time. With the introduction of technologies such as the internet, smartphones, and social media, the pace at which we receive and process information has dramatically increased. In contrast to previous eras, where individuals might have waited days or weeks for news, we now live in a world where information is transmitted almost instantaneously. This acceleration has had a significant impact on how the brain's recursive processes handle time perception.

The Recursive Mind model suggests that the brain's ability to process time is rooted in its recursive loops, which continually refine and adjust temporal information based on sensory input and contextual data. In a digital environment, where individuals are constantly bombarded with notifications, updates, and information streams, these recursive loops are forced to compress time into increasingly smaller units. The result is that time feels as though it is speeding up, days, weeks, or even months seem to pass by more quickly as the brain struggles to keep up with the overwhelming influx of information.

This phenomenon is exacerbated by the fact that digital technology often operates on a different temporal scale than biological processes. While the human body and brain have evolved to function within the constraints of biological rhythms, such as circadian cycles, technology operates on a much faster timescale. Digital devices are capable of processing millions of operations per second, and this creates a disconnect between the brain's natural processing speed and the speed at which information is delivered in the digital age.

As the brain's recursive loops work to synchronise with this accelerated flow of information, time perception becomes increasingly fragmented. The brain is constantly looping through multiple tasks and streams of information, email, text messages, social media feeds, work responsibilities, and personal interactions, leading to a sense of cognitive overload. This overload can distort the perception of time, making it feel as though there is never enough time to keep up with the demands of modern life.

Artificial Intelligence and Time: How AI Changes the Brain's Temporal Interaction

Artificial Intelligence (AI) introduces another layer of complexity to the relationship between technology and time perception. As AI systems become more advanced and integrated into daily life, they are beginning to play a role in how humans navigate, interpret, and experience time. AI operates on a different temporal scale than humans, processing information at speeds far beyond human capabilities, and this raises intriguing questions about how humans and AI might interact in the realm of time.

The Recursive Mind model provides a framework for understanding how humans process time, but AI offers the potential to augment or even disrupt these recursive processes. For example, AI algorithms are now capable of analysing vast amounts of data in real time, allowing them to make predictions and decisions at a speed that humans cannot match. In this context, AI may act as an external recursive processor, helping humans make decisions faster or more efficiently by offloading some of the brain's cognitive load.

One area where this is already occurring is in the realm of personal digital assistants (such as Siri, Alexa, and Google Assistant), which use AI to help individuals manage their time and tasks. These systems can schedule appointments, set reminders, and provide information instantly, effectively reducing the amount of time individuals need to spend on cognitive tasks such as memory recall or decision-making. In doing so, AI allows individuals to compress time even further, potentially altering the brain's natural recursive loops by providing external support for temporal cognition.

However, the introduction of AI into time management also raises important questions about how it might affect the brain's long-term recursive processes. As humans become increasingly reliant on AI for managing time-related tasks, they may offload more of their cognitive responsibilities onto these systems, potentially leading to a reduction in the brain's own recursive capabilities. This phenomenon, sometimes referred to as cognitive outsourcing, could have implications for how the brain processes time, as it may no longer need to engage in the same depth of recursive thinking to manage temporal tasks.

Simulated Time and AI: Can Artificial Systems Experience Time?

One of the most intriguing questions raised by AI is whether artificial systems can experience time in the same way that humans do. Human time perception is deeply tied to consciousness, awareness, and the brain's ability to recursively process sensory input and internal states. AI systems, however, operate without the same biological mechanisms or conscious awareness. Instead, they process data according to pre-programmed algorithms, without any subjective experience of time.

From the perspective of the Recursive Mind model, time perception is not just a result of raw data processing but a product of recursive loops that bind together moments of sensory input, memory, and attention to create a cohesive sense of temporal flow. AI lacks these recursive feedback loops because it does not have internal consciousness or the ability to self-reflect on its experiences. As a result, AI systems do not truly experience time; instead, they operate on a functional level, processing information in a way that mimics human time perception but without any subjective awareness of temporal flow.

However, AI's ability to simulate time can still have profound effects on human time perception. For example, AI systems are increasingly being used to simulate future events, such as in predictive algorithms that forecast everything from weather patterns to stock market trends. These AI-generated simulations are based on recursive data analysis, where the AI continuously loops through vast datasets to generate predictions about future outcomes. In this sense, AI is able to simulate the temporal projection capabilities of the human brain, allowing it to model future scenarios in a way that mirrors human recursive thinking.

While AI may not experience time subjectively, its ability to simulate temporal processes could lead to a new form of cognitive collaboration between humans and machines. By integrating AI into their decision-making processes, humans may be able to enhance their ability to navigate and predict future events, effectively expanding their temporal horizon. This raises important ethical and philosophical questions about the nature of time perception in the age of AI, as well as the potential consequences of relying on artificial systems to augment human cognition.

The Future of Time Perception: How AI and Technology Might Reshape the Recursive Mind

As AI continues to develop and become more integrated into daily life, it is likely that the relationship between humans and time will continue to evolve. One potential future scenario is the development of AI-enhanced brains, where AI systems are directly integrated into human neural networks, creating a hybrid form of cognition that blends biological recursion with artificial processing. In this scenario, AI could potentially augment the brain's recursive loops, allowing individuals to process time-related tasks more efficiently or even expand their temporal awareness.

For example, AI-enhanced brains might allow individuals to simultaneously track multiple timelines, such as managing personal schedules, work tasks, and long-term goals in parallel. The brain's recursive loops would be supplemented by AI's ability to process vast amounts of data in real time, potentially leading to an expanded sense of temporal flow, where individuals could navigate multiple temporal layers at once. This could lead to a form of temporal multitasking that is far beyond the capabilities of the unaugmented human brain.

However, this future also raises important questions about the limits of human cognition and the potential consequences of integrating AI into the brain's recursive loops. While AI may enhance certain aspects of time perception, it could also lead to a disconnection from the more natural rhythms of time, such as biological cycles and the brain's inherent need for rest and recovery. There is a risk that by accelerating time perception to match the speed of AI, humans may become increasingly alienated from the more reflective and contemplative aspects of temporal experience.

Hypothesis: Can AI Reshape the Brain's Recursive Time Processing?

A key hypothesis that emerges from this chapter is that AI has the potential to reshape the brain's recursive processes related to time perception. By offloading cognitive tasks to AI systems, humans may be able to compress time and enhance their ability to navigate complex temporal environments. However, this could also lead to a reduction in the brain's natural recursive capabilities, as individuals become more reliant on external systems to manage temporal tasks.

This hypothesis could be tested by examining how individuals' time perception changes when they use AI systems to manage tasks such as scheduling, decision-making, and planning. By comparing brain activity before and after the use of AI, researchers could explore whether AI alters the brain's recursive loops and whether these changes have a long-term impact on cognitive function and temporal awareness.

The Intersection of Time, Technology, and the Recursive Mind

In conclusion, the rapid development of technology, particularly artificial intelligence, has had a profound impact on how humans perceive and process time. The acceleration of time in the digital age, coupled with the integration of AI into daily life, has created new challenges and opportunities for the brain's recursive processing of time. While AI has the potential to enhance human time perception, it also raises important questions about the future of cognitive interaction with artificial systems and the potential consequences of relying on AI to manage the flow of time.

As we move forward, we will continue to explore how the brain's recursive nature interacts with broader philosophical, biological, and cultural constructs of time, culminating in a deeper understanding of how time shapes human experience across all domains.

Chapter 9: The Biological Foundations of Time Perception: Neural and Cellular Mechanisms

Understanding how the brain processes time requires us to explore its biological foundations, the intricate neural circuits, molecular pathways, and cellular dynamics that underlie time perception. From the firing of neurons to the regulation of synaptic plasticity, time is embedded within the biological machinery of the brain. This chapter delves into the cellular and neurological mechanisms that contribute to our experience of time, examining how neural oscillations, synaptic timing, and molecular clocks work together to create a cohesive sense of temporal flow.

By investigating these biological mechanisms, we will explore how the Recursive Mind model operates at the cellular level, allowing the brain to synchronise different temporal scales, from millisecond timing to long-term memory consolidation. In doing so, we will also examine how disruptions in these mechanisms, whether through disease, injury, or aging, can lead to temporal distortions, providing insight into conditions such as Parkinson's disease, Alzheimer's disease, and other neurological disorders that affect time perception.

Neural Oscillations: The Brain's Internal Clocks and Timekeepers

At the heart of time perception are the brain's neural oscillations, rhythmic patterns of electrical activity that synchronise the firing of neurons across different brain regions. These oscillations occur at various frequencies, ranging from slow delta waves (0.5–4 Hz) associated with deep sleep, to gamma waves (30–100 Hz), which are linked to attention, sensory processing, and consciousness. Each of these oscillatory rhythms plays a critical role in how the brain tracks and processes time, providing the neural framework for temporal binding and synchronisation across different regions of the brain.

Neural oscillations act as the brain's internal timekeepers, creating a temporal scaffold that allows for the precise timing of neural events. In this sense, the oscillatory nature of brain activity functions much like a biological clock, providing a continuous and recursive loop that enables the brain to measure time intervals. For example, in the process of sensory perception, different sensory inputs, such as sight, sound, and touch, arrive at the brain at slightly different times. Neural oscillations help the brain synchronise these inputs, creating a unified perception of the present moment.

The Recursive Mind model suggests that these oscillations are crucial for the recursive loops that underpin temporal cognition. By providing a rhythmic backdrop for neural processing, oscillations allow the brain to loop back to previous moments in time and integrate them with present inputs. This recursive interaction between neural oscillations and cognitive processes enables the brain to maintain a coherent sense of temporal continuity across different time scales.

For example, theta waves (4–8 Hz) are closely associated with memory encoding and retrieval. During tasks that involve recalling past experiences, the brain's theta oscillations synchronise with the hippocampus, the brain region responsible for memory consolidation. This oscillatory activity enables the brain to recursively loop back to past memories, updating them with new information from the present. Similarly, beta oscillations (13–30 Hz) are associated with motor control and timing precision, playing a role in recursive loops that allow the brain to adjust and refine motor actions based on real-time feedback.

Synaptic Timing: Temporal Precision at the Cellular Level

While neural oscillations provide a broad framework for time perception, the timing of individual neural events, particularly at the synaptic level, is equally important for the brain's ability to process time. Synaptic timing refers to the precise moment at which one neuron fires and sends a signal to another neuron across a synapse. This synaptic communication occurs in the span of milliseconds and is critical for tasks that require fine temporal resolution, such as speech processing, music perception, and motor coordination.

At the cellular level, synaptic plasticity, the brain's ability to strengthen or weaken synaptic connections based on activity, plays a crucial role in shaping how the brain encodes time. Long-term potentiation (LTP), a form of synaptic plasticity that strengthens synaptic connections, is believed to underlie learning and memory formation. LTP allows neurons to become more responsive to specific patterns of activity, creating a temporal code that enables the brain to store temporal sequences of events.

The Recursive Mind model suggests that synaptic timing contributes to the brain's recursive loops by enabling the precise coordination of neural events. When neurons fire in a well-timed sequence, they create temporal patterns that can be referenced and looped back to during cognitive processing. For example, during language comprehension, neurons in the brain's auditory cortex fire in synchrony with the rhythmic patterns of speech. These rhythmic patterns are recursively processed by the brain, allowing for the real-time integration of words, sentences, and meaning.

Moreover, the synchrony between different neural circuits, such as those involved in memory, sensory perception, and motor control, relies on the precise timing of synaptic events. Disruptions in synaptic timing, such as those seen in conditions like schizophrenia or autism, can lead to temporal disorganisation, where the brain struggles to synchronise different streams of information, resulting in a fragmented perception of time.

Molecular Clocks and Circadian Rhythms: How Cells Track Time

In addition to the neural circuits responsible for short-term time perception, the brain also relies on molecular clocks to regulate long-term temporal processes, such as the sleep-wake cycle, metabolism, and hormonal regulation. These molecular clocks are governed by circadian rhythms, which are controlled by the suprachiasmatic nucleus (SCN) in the hypothalamus. The SCN acts as the body's master clock, synchronising the activity of molecular clocks in cells throughout the body.

At the molecular level, circadian rhythms are regulated by a network of clock genes and proteins that create self-sustaining feedback loops. These genes and proteins interact in a 24-hour cycle, driving the expression of genes that regulate cellular processes such as energy metabolism, DNA repair, and cell division. This molecular timing system is essential for maintaining the body's internal synchrony with the external environment, allowing the brain and body to anticipate changes in light, temperature, and activity levels.

The Recursive Mind model suggests that these molecular clocks interact with the brain's higher-order cognitive functions, particularly in relation to long-term planning and memory consolidation. For example, the brain's ability to encode and retrieve episodic memories, memories of specific events, depends on the interaction between sleep cycles and molecular timing mechanisms. During sleep, the brain's recursive loops process and consolidate information gathered during the day, integrating it into long-term memory stores.

Disruptions in circadian rhythms, whether through shift work, jet lag, or chronic exposure to artificial light, can interfere with the brain's recursive processes, leading to temporal dysregulation. Individuals who experience circadian misalignment often report feeling disoriented in time, struggling to maintain focus, attention, and memory. This suggests that the synchronisation between molecular clocks and neural circuits is essential for maintaining a coherent sense of time.

Neurodegenerative Disorders and Time Perception: When Timing Breaks Down

Disruptions in the brain's ability to process time are a hallmark of many neurodegenerative disorders, including Parkinson's disease, Alzheimer's disease, and Huntington's disease. These conditions often involve the degeneration of neural circuits that are responsible for temporal precision, leading to significant impairments in time perception, memory, and motor coordination.

For example, individuals with Parkinson's disease often experience difficulty with time estimation and temporal sequencing, particularly in relation to motor tasks. The degeneration of dopaminergic neurons in the basal ganglia, a region of the brain involved in timing and motor control, leads to impairments in the brain's ability to synchronise neural circuits. This results in symptoms such as bradykinesia (slowness of movement), as well as difficulties in anticipating and adjusting to the timing of external events.

Similarly, individuals with Alzheimer's disease often exhibit temporal disorientation, particularly in the later stages of the disease. The degeneration of neurons in the hippocampus and entorhinal cortex, regions critical for memory and spatial navigation, leads to difficulties in recalling the temporal order of events, resulting in confusion about the sequence of past experiences. The brain's recursive loops, which normally allow for the continuous updating and integration of memories, become disrupted, leading to a fragmented sense of time.

These examples highlight the importance of temporal precision at the cellular and neural levels for maintaining a coherent sense of time. When these timing mechanisms break down, whether due to neurodegeneration, injury, or aging, the brain's recursive loops are unable to function effectively, leading to temporal distortions and disorientation.

Hypothesis: Can Targeting Neural Oscillations and Synaptic Timing Improve Time Perception?

A key hypothesis that arises from the Recursive Mind model is that targeting neural oscillations and synaptic timing could improve time perception in individuals with neurological disorders or cognitive impairments. By modulating the brain's oscillatory activity, through techniques such as transcranial magnetic stimulation (TMS) or deep brain stimulation (DBS), it may be possible to restore temporal synchrony and enhance the brain's recursive processing capabilities.

For example, studies have shown that theta stimulation can improve memory encoding and retrieval in individuals with mild cognitive impairment. Similarly, beta stimulation has been used to improve motor timing in individuals with Parkinson's disease. These interventions work by restoring the brain's ability to synchronise neural circuits, allowing for more precise timing of neural events.

By testing how different types of neural stimulation affect time perception and cognitive function, researchers could explore whether enhancing neural oscillations and synaptic timing can improve the brain's ability to process time across different temporal scales. This could have important implications for the treatment of age-related cognitive decline, neurodegenerative diseases, and other conditions that affect time perception.

The Biological Machinery of Time Perception and the Recursive Mind

In conclusion, the biological foundations of time perception, neural oscillations, synaptic timing, and molecular clocks, are essential for the brain's ability to create a coherent sense of time. These mechanisms provide the temporal precision required for the brain's recursive loops to function effectively, allowing us to synchronise past, present, and future into a continuous stream of consciousness.

As we move forward, we will continue to explore how the brain's recursive nature interacts with broader philosophical, cultural, and technological constructs of time, further deepening our understanding of how time shapes human experience at every level of existence.

Chapter 10: Philosophical Theories of Time: Metaphysics, Perception, and the Recursive Mind

The nature of time has been a central question in philosophy for millennia. Philosophers have debated whether time is absolute or relative, whether it exists independently of human perception, or whether it is a mental construct shaped by the way we experience the world. Time, in philosophical thought, has often been seen as a bridge between physical reality and conscious experience, as well as a concept that challenges our understanding of causality, change, and the structure of the universe.

In this chapter, we will explore some of the key philosophical theories of time, focusing on how they intersect with the Recursive Mind model. By examining how the brain's recursive processing of time aligns with various metaphysical ideas, such as presentism, eternalism, and the block universe theory, we will investigate whether the brain's ability to recursively bind together past, present, and future provides insight into these philosophical debates. Additionally, we will explore how philosophical thought on time illuminates the subjective experience of time, and whether the Recursive Mind model can offer new perspectives on temporal perception.

Presentism vs. Eternalism: Is Only the Present Real?

One of the longest-standing debates in the philosophy of time is between presentism and eternalism. Presentism is the view that only the present moment is real, both the past and the future are mere illusions or concepts that exist only in the mind. According to presentism, the present is the only time that truly exists, and our perception of past and future events is merely a reconstruction of memory or anticipation.

In contrast, eternalism argues that the past, present, and future all exist simultaneously. This view is often associated with the block universe theory, which sees time as a fourth dimension, much like space. In this model, all points in time are equally real, and time is more like a frozen block in which past, present, and future coexist. The flow of time, according to eternalism, is a feature of our conscious perception rather than an objective property of the universe.

The Recursive Mind model offers a potential framework for understanding how the brain navigates these philosophical views of time. From the perspective of recursion, the brain continuously loops through past memories, present inputs, and future predictions, creating a seamless experience of temporal flow. In this sense, the brain's recursive loops seem to align more closely with the presentist view, as they focus on constructing a continuous now based on past and future references.

However, the brain's ability to process time recursively also suggests an affinity with eternalism. While we experience time as a linear flow, the brain's recursive loops allow it to access and integrate temporal data from both the past and the future, creating a multidimensional temporal framework. In this way, the brain constructs a mental model of time that incorporates elements of the block universe, where the past and future are accessible at any moment through memory and anticipation.

The Recursive Mind model thus provides a bridge between presentism and eternalism, suggesting that while we experience time as flowing, the brain's ability to recursively navigate temporal dimensions allows it to perceive time in a way that echoes the block universe view. This recursive navigation of time may offer insights into how we conceptualise both the flow of time and the fixed structure of time proposed by eternalism.

The Block Universe and the Experience of Time: A Recursive Interpretation

The block universe theory presents a view of time as a static entity, where the entire timeline of the universe, from its beginning to its end, exists all at once, much like a map that stretches across both space and time. In this view, the passage of time is an illusion created by human perception, and all moments in time are equally real. This raises the question: how do we, as conscious beings, experience the flow of time if the universe itself is a static block?

The Recursive Mind model offers a possible explanation by suggesting that the brain creates the illusion of temporal flow through continuous recursive processing. While the block universe may present time as a fixed dimension, the brain's recursive loops allow us to move through time subjectively, constructing a dynamic experience of past, present, and future. In other words, the brain recursively references different points on the block timeline, creating a narrative of events that gives the impression of time moving forward.

In this sense, the subjective flow of time may be a product of how the brain organises its recursive loops, continuously looping back to past events, integrating them with present sensory input, and projecting forward into the future. The brain's recursive architecture enables it to create a temporal trajectory, allowing us to experience causality and the unfolding of events, even though the block universe model suggests that all moments already exist.

This recursive processing could explain why we experience time as flowing in a particular direction, from past to future, despite the possibility that, at a fundamental level, time itself does not flow. The brain's recursive interaction with temporal information provides the cognitive scaffolding for the arrow of time, allowing us to experience time as moving in a linear fashion even if the underlying reality is more static.

Causality and the Arrow of Time: Recursive Structures in Temporal Perception

Another important philosophical issue surrounding time is the concept of causality, the idea that events in the past cause events in the future. The arrow of time refers to the direction in which time seems to flow, always moving from the past toward the future. This directionality is closely tied to the concept of entropy, as described in the second law of thermodynamics. Entropy, or disorder, tends to increase over time, creating a one-way flow of time from order to disorder.

The Recursive Mind model suggests that the brain's experience of causality and the arrow of time is a product of recursive processes that allow us to track the sequence of events and understand how one event leads to another. The brain builds temporal models by recursively looping through past experiences, allowing us to predict and anticipate future outcomes based on cause-and-effect relationships. This recursive processing gives rise to our experience of temporal asymmetry, where the past influences the future, but not the other way around.

Interestingly, while the brain constructs a unidirectional experience of time, the underlying neural mechanisms are often time-symmetric, meaning that the same neural processes could, in theory, run forward or backward. However, the brain's recursive loops create constraints that give rise to a directional experience of time, aligning with the broader thermodynamic arrow. In this way, the Recursive Mind model bridges the gap between physical laws and subjective experience, showing how the brain's recursive architecture gives rise to our perception of a causal flow of events.

Phenomenology and the Subjective Experience of Time

The phenomenological approach to time, championed by philosophers such as Edmund Husserl and Martin Heidegger, focuses on the lived experience of time rather than its objective structure. According to phenomenology, time is not something that exists independently of human experience; rather, it is deeply intertwined with consciousness and self-awareness. Time, in this view, is something we experience as we live our lives, shaped by our memories, emotions, and intentions.

The Recursive Mind model aligns closely with the phenomenological perspective by emphasising the role of consciousness in creating the experience of time. Through recursive loops, the brain continuously integrates past experiences, present sensory input, and future goals, creating a lived sense of time that is fluid and dynamic. This recursive processing allows the brain to construct a subjective timeline, where moments are connected not by their objective order in the universe but by their emotional and cognitive significance.

For example, when we recall a particularly meaningful event from our past, the brain's recursive loops bring that memory into the present, allowing us to relive the experience in vivid detail. This recursive process creates a temporal fusion of past and present, where the boundaries between then and now become blurred. Similarly, when we imagine a future event, the brain's recursive loops allow us to project ourselves into that future moment, creating a mental simulation that feels as though we are already experiencing it.

This phenomenological approach to time perception highlights the subjective elasticity of time, how time can feel fast or slow, stretched or compressed, depending on our mental state and emotional context. The Recursive Mind model provides a biological framework for understanding this elasticity, showing how the brain's recursive processing of time can expand or contract our experience of temporal intervals based on how we are engaged with the moment.

Time and Free Will: Does Recursive Processing Influence Our Sense of Agency?

The relationship between time and free will has long been a subject of philosophical debate. If the future is already determined, as suggested by deterministic theories of time, then it raises the question of whether we truly have the ability to make free choices that affect the future. On the other hand, if the future is open and undetermined, then how do we navigate a world in which outcomes are not fixed?

The Recursive Mind model offers a perspective on this debate by suggesting that our experience of free will is closely tied to the brain's ability to recursively simulate future possibilities and make decisions based on past experiences. When we engage in decision-making, the brain loops through a series of hypothetical scenarios, evaluating potential outcomes based on prior knowledge and current goals. This recursive evaluation process gives rise to a sense of agency, where we feel as though we are actively shaping the future through our choices.

While the block universe theory suggests that the future already exists and is fixed, the brain's recursive loops allow us to experience the act of choosing among different possibilities, even if those possibilities are already embedded in the fabric of the universe. In this sense, the Recursive Mind model suggests that free will may be an emergent property of recursive processing, giving rise to the subjective feeling of agency even within a deterministic framework.

Hypothesis: Can Recursive Processing Bridge the Gap Between Philosophy and Neuroscience?

A key hypothesis that arises from the Recursive Mind model is that recursive processing in the brain can serve as a bridge between philosophical theories of time and neuroscientific understanding. By examining how the brain's recursive loops enable us to navigate past, present, and future simultaneously, researchers could explore whether the brain's cognitive architecture aligns with different metaphysical views of time, such as presentism, eternalism, and the block universe theory.

This hypothesis could be tested by conducting neuroimaging studies that examine how the brain processes temporal information during tasks that involve memory recall, future planning, and causal reasoning. By comparing brain activity during these tasks, researchers could investigate whether the brain's recursive loops support the idea that the past, present, and future are all processed simultaneously, as suggested by the block universe model, or whether the brain prioritises the present as the only real moment in time, as suggested by presentism.

Philosophical Time and the Recursive Mind

In conclusion, the Recursive Mind model provides a powerful framework for understanding how the brain processes time in ways that intersect with long-standing philosophical debates. By examining how the brain's recursive loops create the experience of temporal flow, causality, and agency, we can gain new insights into whether time is an absolute reality, a mental construct, or something in between.

As we move forward, we will continue to explore how the Recursive Mind model can provide a deeper understanding of time across different domains, from science and philosophy to consciousness and culture, further unravelling the mysteries of how the brain creates meaning from the passage of time.

Chapter 11: Time and Memory: The Reconstruction of Temporal Narratives

Memory is not merely a record of past events but a dynamic process through which the brain reconstructs and reinterprets experiences over time. Time and memory are deeply intertwined; how we perceive the flow of time is influenced by the brain's ability to store and retrieve memories, while our understanding of the past is continually reshaped by present experiences and future expectations. Memory plays a crucial role in the Recursive Mind model, as the brain's recursive loops allow us to revisit past moments, integrate them into the present, and use them to predict future outcomes.

In this chapter, we explore the relationship between time and memory, focusing on how the brain constructs and reconstructs temporal narratives. We will examine how episodic memory, working memory, and semantic memory interact to create a cohesive sense of time, and how the recursive nature of memory allows us to navigate the past, present, and future simultaneously. Additionally, we will explore how false memories and memory distortions arise, shedding light on the flexibility of the brain's temporal processing.

Episodic Memory: The Brain's Temporal Archive

Episodic memory refers to our ability to recall specific events from our personal past. It is the memory system that allows us to mentally travel back in time, re-experiencing moments, people, and places from earlier stages of our lives. Unlike other forms of memory, such as semantic memory (general knowledge) or procedural memory (skills and habits), episodic memory is uniquely tied to the subjective experience of time. When we retrieve an episodic memory, we are not merely recalling a fact or a piece of information; we are reconstructing a temporal narrative, a story of what happened, when it happened, and how it made us feel.

The hippocampus plays a central role in episodic memory, serving as a hub for integrating spatial and temporal information. When we experience an event, the hippocampus encodes not only the details of that event but also the context, the time and place in which the event occurred. This allows us to later retrieve the memory as part of a larger temporal framework, where past experiences are organised in relation to one another.

The Recursive Mind model suggests that episodic memory operates through recursive loops that allow the brain to revisit and update past experiences. Each time we recall a memory, the brain loops back to the original event, integrating it with new information from the present. This recursive process means that episodic memories are not static; they are constantly evolving, shaped by our current thoughts, feelings, and interpretations.

For example, if you recall a childhood memory of a family vacation, your brain does not simply retrieve an exact copy of that event. Instead, it reconstructs the memory by looping through different sensory details (the sound of the ocean, the smell of sunscreen) and emotional reactions (joy, excitement, or nostalgia). This reconstruction is influenced by your current mental state, how you feel about your family now, how you interpret that vacation in light of your adult experiences, and how the passage of time has altered your perspective on that event.

This recursive process of memory reconstruction is what allows us to create a coherent temporal narrative of our lives. By continuously revisiting and updating past experiences, the brain integrates them into a broader story of who we are, how we have changed, and where we are headed. This narrative function of memory is essential for maintaining a sense of self across time, as it allows us to see ourselves as the same person who experienced the past, lives in the present, and will continue into the future.

Working Memory: The Present Moment in Focus

While episodic memory allows us to mentally travel back in time, working memory is responsible for keeping information in mind in the present moment. Working memory acts as a mental workspace, where the brain temporarily holds and manipulates information needed for ongoing tasks. Whether you are solving a math problem, holding a phone number in mind, or following the steps of a recipe, working memory allows you to maintain focus on the immediate task at hand.

The prefrontal cortex is the brain region most closely associated with working memory. It acts as a gatekeeper, determining which pieces of information are relevant to the current task and should be held in mind. Working memory is highly dynamic, with information being continuously updated, discarded, or replaced based on the demands of the present moment.

In the context of the Recursive Mind model, working memory plays a key role in the brain's ability to process time because it allows us to sustain attention across short time intervals. By looping back to recently processed information, working memory enables the brain to create a sense of temporal continuity. For example, when you are engaged in a conversation, your working memory holds the words and ideas that were just spoken, allowing you to respond in a coherent and timely manner. This process of looping back to recent information creates a seamless flow of temporal integration, where each moment is connected to the one before it.

Working memory is also essential for planning and decision-making, as it allows the brain to hold multiple pieces of information in mind while considering future outcomes. When you plan your day, for example, your working memory loops through a series of hypothetical scenarios, evaluating how different choices will affect your future schedule. This recursive process allows the brain to simulate future events based on current information, enabling you to make decisions that are grounded in both the present and the future.

However, working memory has its limits. It can only hold a limited amount of information at a time, typically around seven items, and its capacity can be easily overloaded by distractions or competing demands. When working memory is overloaded, the brain's ability to maintain temporal coherence is compromised, leading to difficulties in tracking time or staying focused on the task at hand.

Semantic Memory and Temporal Generalisation

While episodic memory is tied to specific events in time, semantic memory is concerned with general knowledge, the facts, concepts, and meanings that we accumulate over a lifetime. Semantic memory is not tied to a particular moment in time but instead represents timeless knowledge that can be accessed and applied whenever needed. For example, knowing that the Earth revolves around the Sun or that Paris is the capital of France are examples of semantic memory.

The Recursive Mind model suggests that semantic memory plays a role in shaping our temporal perception by providing a framework for generalising time. While episodic memory is focused on specific moments, semantic memory allows us to make broad generalisations about time, such as understanding the concepts of yesterday, today, and tomorrow, or knowing how long a typical year lasts.

Semantic memory interacts with the brain's recursive loops by providing temporal anchors that help us organise our experiences into a coherent timeline. For example, when you recall a historical event, such as the signing of the Declaration of Independence, your brain uses semantic memory to place that event within a broader temporal framework. You may not have personally experienced the event, but your knowledge of history allows you to understand when and where it occurred relative to other events.

This interaction between episodic and semantic memory is essential for creating a comprehensive sense of time. While episodic memory allows us to recall specific events from our personal past, semantic memory provides the broader context that enables us to situate those events within the larger narrative of history, culture, and the universe.

False Memories and Temporal Distortions: The Flexibility of Memory Reconstruction

One of the most fascinating aspects of memory is its flexibility. Because memory is a reconstructive process, it is subject to distortions and inaccuracies. This is particularly true when it comes to false memories, memories of events that never actually occurred. False memories can arise from a variety of factors, including suggestion, imagination, and the brain's tendency to fill in gaps in memory with plausible details.

The Recursive Mind model offers a potential explanation for how false memories arise. Because the brain's recursive loops continuously revisit and update past experiences, memories are constantly being reconstructed based on new information. Each time we recall a memory, the brain may introduce subtle modifications, incorporating new details or altering the context in which the event occurred. Over time, these modifications can accumulate, leading to the creation of a false memory that feels just as real as a true memory.

False memories often involve temporal distortions, where the brain misremembers the timing of an event or places it in the wrong sequence. For example, you might recall meeting a friend at a specific party, only to later discover that the friend wasn't there. The brain's recursive loops may have mistakenly merged two separate events, creating a false memory that blends details from both. This ability to reconstruct time allows the brain to create a flexible and adaptive sense of the past, but it also makes memory vulnerable to temporal errors.

Memory, Time, and the Self: Constructing Temporal Identity

At the heart of memory's relationship with time is the role it plays in constructing our sense of self. Our memories are not just a collection of past events; they are the building blocks of our personal identity, allowing us to understand who we are and how we have changed over time. The Recursive Mind model suggests that memory is essential for creating a temporal identity, a coherent narrative that connects the past, present, and future into a unified sense of self.

Through recursive processing, the brain loops through past memories and future projections, allowing us to create a temporal continuity that gives meaning to our experiences. This continuity is essential for maintaining a sense of agency and purpose, as it allows us to see our actions as part of a larger story that stretches across time.

For example, when you reflect on a significant life decision, such as choosing a career or starting a family, you are not simply recalling an isolated event. Instead, you are engaging in a recursive process that integrates past choices, present circumstances, and future goals into a coherent narrative. This narrative function of memory is what allows us to create a sense of temporal direction, where the past informs the present, and the present shapes the future.

Hypothesis: Can Enhancing Recursive Memory Processing Improve Temporal Awareness?

A key hypothesis that arises from the Recursive Mind model is that enhancing the brain's ability to process memory recursively could improve temporal awareness and time perception. By strengthening the connections between episodic memory, working memory, and semantic memory, it may be possible to improve the brain's ability to track time, maintain temporal coherence, and avoid distortions such as false memories.

This hypothesis could be tested by developing cognitive training programs or neurostimulation techniques that target the brain's recursive memory circuits. By enhancing the brain's ability to loop through past experiences and integrate them with present information, researchers could explore whether these interventions lead to improvements in memory accuracy, temporal orientation, and overall time management.

Memory as a Time Machine in the Recursive Mind

In conclusion, memory is not just a record of the past but a dynamic process that allows the brain to navigate time. Through recursive loops, the brain continuously reconstructs and updates past experiences, integrating them into a coherent temporal narrative that shapes our sense of self and our understanding of time. The flexibility of memory allows us to revisit the past, live in the present, and plan for the future, all while maintaining a continuous sense of temporal flow.

As we move forward, we will continue to explore how the brain's recursive nature interacts with other aspects of time perception, such as consciousness, emotion, and cultural constructs, further deepening our understanding of how the brain creates meaning from the passage of time.

Chapter 12: Emotion and Time Perception, The Recursive Interplay Between Feelings and Temporal Experience

Time perception is not only a cognitive process but also deeply intertwined with our emotional states. Emotions have the power to expand, contract, or distort our sense of time, depending on the intensity and nature of the feeling. Whether time seems to drag during moments of boredom or speed up during periods of excitement, our emotional experience significantly shapes how we perceive the passage of time. This chapter delves into the complex relationship between emotion and time perception, exploring how the brain's recursive loops modulate the interaction between our feelings and our temporal awareness.

We will examine how different emotional states, such as joy, fear, anxiety, and sadness, alter our perception of time, and how the brain processes these experiences through recursive feedback loops that integrate past emotional experiences with present sensory input. Additionally, we will explore the neurological mechanisms behind the emotional modulation of time perception, focusing on the roles of the amygdala, prefrontal cortex, and limbic system. By understanding how emotions influence our sense of time, we can gain insights into both normal and abnormal time perception in conditions such as depression, post-traumatic stress disorder (PTSD), and anxiety disorders.

The Elasticity of Time: How Emotion Shapes Temporal Perception

One of the most striking ways that emotion influences time perception is through its ability to make time feel elastic, stretching or compressing depending on our emotional state. When we are engaged in positive experiences, such as a joyful social gathering or an exciting event, time often seems to fly by. In contrast, during moments of negative emotion, such as boredom, anxiety, or sadness, time can feel as though it is dragging, with each second stretching out endlessly.

This elasticity in time perception is a direct result of how the brain processes emotion through recursive loops. When we experience a heightened emotional state, the brain enters a feedback loop where it continuously revisits the emotional intensity of the moment, amplifying the experience and affecting how we perceive the passage of time. For example, during moments of fear or anxiety, the brain's amygdala, the region responsible for processing fear responses, activates recursive loops that heighten vigilance and alertness. This heightened state of awareness causes time to feel stretched, as the brain devotes more resources to monitoring the environment and preparing for potential threats.

Conversely, during periods of happiness or engagement, the brain's reward circuits, including the nucleus accumbens and dopamine pathways, become highly active, leading to a compression of time perception. In these moments, the brain's recursive loops focus on sustaining the positive emotional experience, filtering out distractions and compressing the sense of time to maximise the present moment.

These emotional modulations of time perception can be understood through the lens of the Recursive Mind model, where the brain's ability to loop through emotional and sensory information creates a dynamic and flexible sense of temporal flow. The recursive nature of emotional processing allows the brain to constantly update its perception of time based on the current emotional context, leading to the subjective distortion of time in both positive and negative emotional states.

Fear, Anxiety, and the Slowing of Time

Fear and anxiety are two of the most powerful emotional states that can dramatically slow down our perception of time. When we encounter a threatening situation, the brain's fight-or-flight response is activated, leading to a heightened sense of vigilance and awareness. This response is primarily mediated by the amygdala, which triggers a cascade of neural signals that prepare the body to respond to danger.

From the perspective of the Recursive Mind model, the brain's response to fear involves a recursive feedback loop in which the brain continuously revisits the emotional intensity of the threat. The amygdala sends signals to the prefrontal cortex and other regions of the brain, amplifying the perception of time by increasing the brain's focus on immediate sensory inputs. This recursive process allows the brain to process and analyse every detail of the environment, preparing for a rapid response to the perceived danger.

The result is a sense that time has slowed down, as the brain dedicates more resources to processing sensory information and potential outcomes. This phenomenon, known as time dilation, is commonly experienced during moments of acute fear, such as during a car accident or life-threatening situation. The brain's recursive loops heighten awareness of each passing moment, allowing individuals to react more quickly to the danger but also creating the subjective experience of elongated time.

Anxiety, while less intense than fear, similarly affects time perception by increasing cognitive load and worry about future events. Individuals with generalised anxiety disorder often report that time seems to pass more slowly when they are ruminating on uncertainties or potential threats. In these cases, the brain's recursive loops continuously revisit the anxious thoughts, keeping the individual in a heightened state of anticipation. This recursive processing of anxiety can distort the perception of time, making it feel as though time is moving more slowly than it actually is.

Joy, Excitement, and the Compression of Time

In contrast to fear and anxiety, positive emotions, such as joy, excitement, and engagement, tend to compress time, making it feel as though it is moving more quickly. This is because the brain's reward circuits are highly active during positive emotional experiences, focusing attention on the present moment and filtering out distractions that might otherwise slow down time perception.

When we are engaged in a joyful activity, such as spending time with loved ones or pursuing a passion, the brain's dopamine system creates a recursive loop that reinforces the positive emotional experience. The brain continuously loops through sensory feedback, such as the sound of laughter, the warmth of touch, or the excitement of a new discovery, creating a self-sustaining cycle of positive reinforcement. As the brain becomes more absorbed in the present moment, it filters out irrelevant information, leading to the compression of time.

The Recursive Mind model suggests that the brain's ability to narrow attention and sustain focus during positive emotional states contributes to the subjective experience of time speeding up. By recursively looping through the rewarding aspects of the experience, the brain creates a sense of temporal flow where each moment feels connected to the next, but time seems to pass more quickly because we are fully engaged.

This compression of time during positive emotional states is often referred to as being "in the zone" or experiencing flow. Flow states occur when individuals are deeply immersed in an activity that challenges them just enough to maintain focus but not so much that it induces stress. During flow, the brain's recursive loops efficiently process the activity, creating a sense of effortless action and time compression.

Sadness and Time Expansion: The Weight of Emotional Heaviness

Sadness, particularly in its more extreme forms such as grief and depression, can have the opposite effect on time perception, making time feel heavy and stretched. Individuals experiencing deep sadness often report that time seems to move painfully slowly, with each minute dragging on as they endure the emotional weight of their experience.

The Recursive Mind model suggests that this slowing of time during sadness is due to the brain's tendency to loop through and ruminate on the emotional pain. When we experience sadness, the brain's recursive loops continuously revisit the emotional source of the pain, reinforcing the negative feelings and preventing the individual from moving on from the experience. This recursive processing of sadness can trap individuals in a cycle of rumination, where they repeatedly focus on the loss or disappointment, making time feel as though it is standing still.

For example, individuals who are grieving the loss of a loved one often experience time distortion, where it feels as though time has come to a halt, and the world has stopped moving. This emotional heaviness can make it difficult for individuals to engage with the present moment, as their brain is constantly looping through the past experience of the loss, preventing them from moving forward.

In cases of clinical depression, the brain's recursive loops may become dysregulated, leading to a persistent focus on negative emotions and thoughts. This can result in a chronic distortion of time perception, where the individual feels as though they are stuck in an endless loop of sadness, with time moving slowly and painfully.

Emotion and Memory: The Role of Recursion in Emotional Time Recall

Emotion not only affects our immediate perception of time but also plays a significant role in how we remember past experiences. Emotional memories tend to be more vivid and long-lasting than neutral memories, and this emotional intensity can distort how we recall the timing of past events. For example, we are more likely to remember exactly when a traumatic or joyful event occurred, while less emotionally significant events may become blurred in our memory.

The Recursive Mind model suggests that emotional memories are subject to recursive reprocessing, where the brain continuously loops through the emotional intensity of the past experience. Each time we recall an emotional memory, the brain revisits the emotional context in which the event occurred, amplifying its significance and potentially altering our perception of when it took place. This recursive loop allows emotional memories to be reconstructed over time, with the emotional intensity of the memory influencing how we place it within our temporal framework.

For instance, traumatic memories often feel as though they happened more recently than they actually did. This is because the brain's recursive loops continue to replay the emotional intensity of the trauma, keeping the memory at the forefront of consciousness. Even if the event occurred years ago, the emotional weight of the memory can make it feel immediate, as though it is still affecting the individual in the present moment. This phenomenon is common in individuals with post-traumatic stress disorder (PTSD), where traumatic memories intrude into daily life, distorting the perception of both past and present time.

Conversely, positive emotional memories can also create a temporal distortion, though in a different way. Happy memories, such as a wedding day, the birth of a child, or a personal achievement, are often remembered as timeless or fleeting. The recursive reprocessing of these memories amplifies the positive emotions associated with them, and individuals often recall these events as if they occurred in a magical moment outside the normal flow of time. In these cases, the emotional intensity compresses the memory, creating a sense that the moment passed too quickly despite its vividness in memory.

The Recursive Mind model highlights the flexibility of memory in relation to time, showing that emotional memories are not simply stored as objective records but are continuously reshaped and revisited through recursive loops. These loops create a dynamic interaction between emotion and temporal perception, where the emotional weight of a memory influences how we perceive its place in time.

The Brain's Emotional Timekeepers: Neurological Mechanisms of Emotion and Time Perception

The relationship between emotion and time perception is rooted in the neurological circuits that process both emotional responses and temporal information. Several key brain regions are involved in this recursive interplay, including the amygdala, the prefrontal cortex, the insula, and the anterior cingulate cortex. These regions work together to modulate how we experience time during emotional events and how we recall emotional memories later on.

- Amygdala: The amygdala is central to processing fear, anxiety, and emotional arousal. It plays a critical role in the recursive loops that heighten awareness during threatening situations, leading to time dilation. The amygdala's interactions with the prefrontal cortex allow the brain to integrate emotional context into its temporal processing, making threatening events feel longer or more drawn out.
- Prefrontal Cortex: The prefrontal cortex is involved in executive functions, such as decision-making, attention, and working memory, and plays a key role in emotion regulation. It helps modulate the recursive loops between emotional and temporal processing by directing attention to relevant emotional stimuli or filtering out irrelevant information. During positive emotional states, the prefrontal cortex helps sustain focus on rewarding activities, leading to time compression.
- Insula: The insula is associated with self-awareness, bodily sensations, and emotional experience. It integrates internal body signals (such as heart rate, respiration, and emotional arousal) with the brain's temporal processing, contributing to the subjective experience of time during emotionally charged moments. The insula's recursive processing of bodily and emotional signals allows individuals to feel more intensely connected to the passage of time during moments of heightened emotion.
- Anterior Cingulate Cortex (ACC): The ACC is involved in error detection, conflict monitoring, and emotional

regulation. It plays a role in the recursive loops that evaluate the emotional significance of events, helping the brain to assess whether a situation is rewarding or threatening. The ACC's interactions with the amygdala and prefrontal cortex contribute to the modulation of time perception based on emotional valence.

Together, these brain regions create a recursive feedback network that modulates how we experience time based on our emotional state. Whether we are feeling joy, fear, sadness, or excitement, the brain's recursive loops integrate emotional and temporal information, shaping our subjective experience of the flow of time.

Emotion, Time Perception, and Mental Health: Disorders of Temporal Processing

Disruptions in the brain's recursive processing of emotion and time perception can lead to distorted temporal experiences, particularly in individuals with mental health disorders. For example, individuals with depression often report feeling as though time is moving painfully slowly or that they are stuck in an endless loop of sadness. This temporal distortion is a result of the brain's recursive loops continuously revisiting the emotional pain, preventing the individual from moving forward in time.

Similarly, individuals with PTSD experience temporal flashbacks, where traumatic memories intrude into the present and disrupt the normal flow of time. These flashbacks are driven by the brain's recursive loops replaying the trauma, causing the individual to feel as though they are reliving the event rather than simply recalling it. The result is a profound distortion of time, where the past feels as though it is bleeding into the present.

Conversely, individuals with mania or hypomania, as seen in bipolar disorder, may experience the opposite distortion, where time seems to speed up and they feel as though they are moving at an accelerated pace. This compression of time is often accompanied by racing thoughts, increased energy, and distractibility, as the brain's recursive loops struggle to keep up with the heightened emotional and cognitive activity.

Understanding the recursive interplay between emotion and time perception offers valuable insights into the treatment of mental health disorders. By targeting the brain's recursive loops, whether through cognitive-behavioural therapy (CBT), mindfulness, or pharmacological interventions, it may be possible to help individuals regain a more balanced and adaptive sense of time, reducing the emotional and temporal distortions that contribute to their suffering.

Hypothesis: Can Emotion Modulation Training Improve Temporal Perception?

A key hypothesis that emerges from the Recursive Mind model is that emotion modulation training, such as mindfulness meditation or emotion regulation strategies, could improve temporal perception by enhancing the brain's ability to balance emotional and temporal processing. By training individuals to regulate their emotional responses and focus their attention on the present moment, it may be possible to reduce the temporal distortions caused by intense emotional states.

This hypothesis could be tested by examining how mindfulness training or emotion regulation therapy affects time perception in individuals with anxiety, depression, or PTSD. By measuring changes in brain activity, emotional regulation, and subjective time perception before and after training, researchers could explore whether these interventions improve the brain's ability to process time more accurately and flexibly.

The Recursive Interplay of Emotion and Time Perception

In conclusion, emotion plays a central role in shaping how we experience time. Through recursive processing, the brain integrates emotional intensity with temporal information, creating subjective distortions of time based on our feelings in the moment. Whether time feels fast, slow, heavy, or fleeting, our emotions influence how we perceive its passage, and the recursive nature of emotional processing allows us to constantly update and reshape our temporal experiences.

As we move forward, we will continue to explore how the brain's recursive nature interacts with other dimensions of time perception, such as consciousness, cultural constructs, and philosophical perspectives, further deepening our understanding of the intricate relationship between time and the human mind.

Chapter 13: Cultural and Societal Constructs of Time: The Social Frameworks of Temporal Experience

Time is not only a biological or psychological phenomenon but also a deeply social construct. Different cultures, societies, and historical periods have vastly different ways of understanding, organising, and valuing time. These cultural constructs shape not only how individuals experience time but also how they interact with others, plan their lives, and navigate their roles within society. This chapter explores how cultural norms, societal expectations, and historical contexts influence the perception of time and how the Recursive Mind model interacts with these external frameworks to create a cohesive temporal experience.

We will delve into how cultures vary in their views of time, whether it is seen as linear or cyclical, monochronic or polychronic, and how these perceptions influence both individual cognition and social dynamics. Additionally, we will investigate how modern technology, globalisation, and social media have impacted time perception, accelerating the pace of life in many parts of the world and changing the way people manage and measure time.

Linear vs. Cyclical Time: How Cultures Perceive the Flow of Time

One of the most fundamental distinctions in cultural conceptions of time is the difference between linear and cyclical views of time. In many Western cultures, time is predominantly understood as linear, a straight line that moves from the past through the present and into the future. This perspective emphasises the irreversibility of time, with a clear focus on progress, planning, and moving toward the future. In such cultures, time is often perceived as a scarce resource, something that must be used efficiently and productively.

In contrast, many Eastern and indigenous cultures view time as cyclical, a repeating loop where events occur in cycles, such as the seasons, the phases of the moon, or the cycles of birth, death, and rebirth. In this view, time is less about linear progress and more about maintaining harmony and balance within the cycles of nature. The past, present, and future are often seen as interconnected, with events from the past recurring in the present or future in different forms.

The Recursive Mind model offers a useful lens for understanding how these different cultural views of time are internalised by individuals. In cultures that emphasise linear time, the brain's recursive loops may become oriented toward planning and goal-setting, with a strong focus on future outcomes. These recursive loops continuously process past experiences in relation to current goals, allowing the individual to anticipate and navigate future events with a forward-looking mindset.

In cultures that embrace a cyclical view of time, the brain's recursive loops may focus more on pattern recognition and repetition, with an emphasis on the recurrence of events rather than linear progression. The brain may become attuned to recognising natural cycles and integrating past experiences into these recurring patterns, creating a sense of continuity across time. This recursive looping through cyclical experiences allows individuals to engage with time in a way that emphasises balance and interconnection rather than progress or change.

Monochronic vs. Polychronic Time: Social Rhythms and Temporal Organisation

Another important cultural distinction is between monochronic and polychronic approaches to time. In monochronic cultures, such as those found in the United States, Germany, and much of Northern Europe, time is seen as discrete and sequential, with a strong emphasis on schedules, punctuality, and doing one thing at a time. In these cultures, time is highly organised and structured, and people are expected to adhere strictly to appointments and deadlines. Time is treated as a commodity that can be measured, saved, or wasted.

In contrast, polychronic cultures, such as those in many parts of Latin America, the Middle East, and Africa, have a more fluid and flexible approach to time. People in polychronic cultures often engage in multiple activities simultaneously and place a higher value on relationships and social interactions than on strict adherence to schedules. Time is viewed as abundant, and it is more important to be adaptable and responsive to changing circumstances than to rigidly follow a predetermined plan.

The Recursive Mind model suggests that these cultural differences in time organisation influence how individuals structure their mental processes. In monochronic cultures, the brain's recursive loops may become highly focused on sequencing tasks and prioritising goals in a linear fashion, with a strong emphasis on efficiency and completion. Individuals in these cultures may experience time as something to be managed carefully, and their recursive processing may reflect a continuous loop of task-oriented thinking.

In polychronic cultures, the brain's recursive loops may be more multidimensional, allowing individuals to engage in multiple tasks or social interactions simultaneously. Rather than looping through a single, linear sequence of tasks, the brain may shift fluidly between different activities and contexts, creating a sense of temporal flexibility. This recursive processing allows individuals to prioritise social relationships and adapt to changes in their environment without feeling bound by strict schedules.

Social Timekeeping: Calendars, Clocks, and Temporal Norms

The way societies measure and organise time plays a significant role in shaping how individuals experience it. Calendars, clocks, and time zones are tools that allow societies to impose a shared temporal structure on their members, creating a common framework for coordinating activities and regulating daily life. These social timekeeping systems reflect deeper cultural values and historical developments, influencing everything from religious practices to economic productivity.

For example, the development of the Gregorian calendar in the West was motivated by both religious and agricultural needs, and its widespread adoption helped standardise time across different regions. Similarly, the invention of mechanical clocks during the Industrial Revolution allowed societies to regulate work hours, creating a more punctual and precise relationship with time. These timekeeping tools transformed the way people organised their lives, shifting from a more natural relationship with time (based on the sun, moon, and seasons) to a more mechanised and standardised one.

In the modern era, digital technology and globalisation have further transformed how societies organise time. The widespread use of smartphones, digital calendars, and social media has created a sense of immediacy and instant connectivity, where individuals are constantly aware of the time and expected to respond to information and tasks in real-time. This acceleration of time has had a profound impact on how individuals perceive the pace of life and manage their time in both personal and professional contexts.

The Recursive Mind model suggests that the brain's recursive loops adapt to these social timekeeping systems, aligning cognitive processes with the broader temporal norms of the culture. For example, in societies where punctuality and strict schedules are valued, the brain's recursive loops may become oriented toward monitoring time closely, allowing individuals to track appointments and deadlines with precision. In more flexible cultures, the brain's recursive loops may prioritise social engagement and adaptability, creating a more relaxed relationship with time.

Technological Time Compression: The Impact of the Digital Age

With the rise of digital technology, the pace of life in many parts of the world has accelerated dramatically. The internet, smartphones, and social media have created a world where information is available instantly, and where people are expected to respond to messages, emails, and notifications in real-time. This compression of time has led to a sense of time scarcity, where individuals feel as though there is never enough time to complete tasks or fully engage with the present moment.

The Recursive Mind model offers insight into how digital technology affects the brain's processing of time. In a digital environment, the brain's recursive loops are constantly being interrupted by new stimuli, emails, notifications, and social media updates. Each time a new piece of information is introduced, the brain must loop back to assess its relevance, prioritise it, and integrate it into the ongoing cognitive process. This continuous looping creates a sense of cognitive overload, where the brain is forced to manage multiple streams of information simultaneously.

As a result, time may feel as though it is moving faster, as the brain struggles to keep up with the demands of constant connectivity and information flow. This acceleration of time can lead to feelings of stress, anxiety, and burnout, as individuals feel pressured to respond to information and tasks as quickly as possible. The recursive loops that normally allow the brain to process time in a balanced and adaptive way may become overloaded, leading to a fragmented sense of time where the present moment feels compressed and fleeting.

However, the recursive nature of the brain also allows for adaptation. Over time, individuals may develop strategies for managing the accelerated pace of digital life, learning to filter out irrelevant information and focus on what is most important. By training the brain's recursive loops to prioritise focus and attention, individuals may be able to regain a sense of temporal balance in the midst of the digital age.

Globalisation and Time Standardisation: Synchronising the World

The modern world is increasingly connected by global networks, where people in different time zones and cultural contexts interact on a daily basis. This globalisation of time has created a need for temporal standardisation, where individuals must coordinate activities across different regions, each with its own local time. The invention of time zones in the 19th century, and their subsequent adoption across the globe, allowed for the synchronisation of activities such as international trade, travel, and communication.

However, the standardisation of time has also created new challenges for individuals and societies. The need to coordinate activities across different time zones can lead to temporal dissonance, where people must adjust their schedules to align with global norms. This is particularly evident in the world of business and finance, where people often work across multiple time zones, resulting in irregular work hours and jet lag.

The Recursive Mind model suggests that the brain's recursive loops must constantly adjust to these global time demands, reorienting cognitive processes to align with the external environment. For individuals who regularly work across time zones, the brain's recursive loops may become attuned to managing multiple temporal frameworks simultaneously. This ability to navigate different time zones requires a high level of cognitive flexibility, as the brain must loop back and forth between local and global time, integrating these different temporal perspectives into a cohesive sense of time.

Hypothesis: Can Recursive Processing Help Manage Temporal Acceleration in the Digital Age?

A key hypothesis that emerges from the Recursive Mind model is that training the brain's recursive loops to manage the accelerated pace of digital life could improve time perception and reduce feelings of stress and overload. By developing strategies to filter out irrelevant information and prioritise focus on the most important tasks, individuals may be able to regain a sense of temporal balance and reduce the cognitive overload associated with digital technology.

This hypothesis could be tested by developing cognitive training programs that focus on enhancing the brain's ability to manage multiple streams of information simultaneously. By measuring changes in brain activity, time perception, and emotional well-being before and after training, researchers could explore whether these interventions help individuals adapt to the accelerated pace of digital life and regain a more balanced relationship with time.

The Social Construction of Time and the Recursive Mind

In conclusion, time is not only a biological or psychological phenomenon but also a deeply social construct that shapes how individuals experience the world. Different cultures and societies have vastly different ways of organising and valuing time, and these social frameworks influence both individual cognition and social dynamics. The Recursive Mind model provides a powerful framework for understanding how the brain's recursive loops interact with these cultural constructs, allowing individuals to navigate the complex relationship between personal and social time.

As we move forward, we will continue to explore how the brain's recursive nature interacts with other aspects of time perception, such as philosophical theories and biological rhythms, further deepening our understanding of how time shapes human experience at every level.

Chapter 14: Cultural and Societal Constructs of Time: The Social Frameworks of Temporal Experience

Time perception is deeply influenced by the body's biological rhythms, which provide the underlying framework for how we experience the passage of time. These rhythms, such as the circadian cycle, ultradian rhythms, and infradian rhythms, regulate everything from sleep-wake cycles to hormone release and cellular metabolism. The brain's ability to synchronise these internal clocks with the external environment is essential for maintaining a coherent sense of time, allowing individuals to align their cognitive processes with the natural flow of the day and night, seasons, and even life stages.

In this chapter, we will explore how the brain's internal clocks interact with external time cues (such as light and temperature), and how disruptions to these biological rhythms, whether through jet lag, shift work, or chronic sleep deprivation, affect time perception. We will also examine how the Recursive Mind model operates at the level of biological rhythms, showing how the brain's recursive loops integrate temporal information from both internal and external sources to create a dynamic and flexible sense of time.

The Circadian Rhythm: The Body's Master Clock

The circadian rhythm is the body's primary timekeeping system, governing the 24-hour cycle of physiological processes such as sleep, wakefulness, body temperature, and hormone levels. The circadian rhythm is regulated by the suprachiasmatic nucleus (SCN), a small cluster of neurons in the hypothalamus that acts as the brain's master clock. The SCN synchronises the body's internal clocks with external time cues, such as light and darkness, ensuring that physiological processes are aligned with the external environment.

The Recursive Mind model suggests that the brain's recursive loops play a crucial role in maintaining the synchronisation between internal and external time. The SCN continuously loops through sensory input, particularly from the retina, to monitor changes in light levels and adjust the body's internal clocks accordingly. This recursive process allows the brain to maintain a sense of temporal coherence across different times of day, ensuring that cognitive and physiological processes are optimally timed for the individual's environment.

For example, the release of the hormone melatonin, which helps regulate sleep, is triggered by the onset of darkness and suppressed by exposure to light. The SCN sends signals to the pineal gland, which produces melatonin, helping the body prepare for sleep at the appropriate time. This process relies on the recursive integration of external light cues with the body's internal timekeeping mechanisms, allowing the individual to maintain a regular sleep-wake cycle.

Disruptions to the circadian rhythm, such as those caused by jet lag or shift work, can interfere with this synchronisation, leading to temporal disorientation and a distorted perception of time. For example, individuals who travel across multiple time zones often experience jet lag, where the body's internal clock is out of sync with the local time. This desynchronisation can result in feelings of fatigue, cognitive impairment, and a sense that time is either moving too slowly or too quickly, depending on the degree of misalignment.

Ultradian and Infradian Rhythms: Beyond the 24-Hour Cycle

While the circadian rhythm governs the body's 24-hour cycle, other biological rhythms operate on different timescales. Ultradian rhythms are shorter cycles, typically lasting less than 24 hours, and include processes such as the 90-minute sleep cycle, hormonal fluctuations, and the digestive process. Infradian rhythms, on the other hand, are longer cycles that extend beyond 24 hours, such as the menstrual cycle or the seasonal rhythms that govern reproductive cycles and metabolic changes in certain animals.

The brain's recursive loops allow it to integrate these multiple temporal scales, ensuring that cognitive and physiological processes are coordinated across both short-term and long-term cycles. For example, during sleep, the brain cycles through different stages, light sleep, deep sleep, and REM sleep, in a roughly 90-minute ultradian rhythm. The brain's recursive loops monitor the transitions between these stages, ensuring that the body progresses through each stage in the proper sequence.

Similarly, the brain's recursive processing of infradian rhythms helps regulate processes that occur over longer time periods, such as the hormonal changes associated with the menstrual cycle. These longer cycles are integrated into the brain's temporal framework, allowing individuals to maintain a sense of temporal coherence even as their bodies undergo changes over weeks or months.

The Recursive Mind model highlights the brain's ability to process temporal information from both internal and external sources simultaneously. By looping through the various biological rhythms that govern bodily processes, the brain is able to create a multi-layered sense of time, where short-term and long-term cycles are seamlessly integrated into a coherent experience of temporal flow.

Biological Time and Aging: The Impact of Time on the Body

As we age, our biological rhythms change, affecting how we perceive and experience time. The aging process involves the gradual decline of the body's ability to maintain the precise timing of its internal clocks, leading to disruptions in circadian rhythms and other biological cycles. These changes can have a profound impact on cognitive function, sleep patterns, and overall well-being.

For example, older adults often experience a shift in their circadian rhythm, becoming morning types (or "larks") who wake up and go to bed earlier than they did in their younger years. This shift in circadian timing is associated with a decline in melatonin production and a reduced sensitivity to light cues, which can lead to sleep disturbances and a fragmented sense of time.

The Recursive Mind model suggests that the brain's recursive loops may become less efficient at synchronising internal and external time cues as we age, leading to a less precise sense of time. This decline in recursive processing may contribute to the experience of temporal disorientation or time distortion in older adults, where days, weeks, or even months seem to blur together. Additionally, the loss of neuronal plasticity, the brain's ability to form and strengthen connections between neurons, may reduce the brain's capacity to adapt to changes in its biological rhythms, further contributing to time-related cognitive impairments.

In some cases, the aging process may also lead to neurodegenerative diseases such as Alzheimer's disease or Parkinson's disease, which are characterised by significant disruptions in time perception and biological rhythms. For example, individuals with Alzheimer's disease often experience circadian misalignment, where their sleep-wake cycles become irregular, leading to confusion and disorientation about the time of day.

Seasonal and Environmental Influences on Time Perception

In addition to the body's internal clocks, time perception is also influenced by external environmental factors, such as seasonal changes, temperature, and daylight hours. For example, during the winter months, when daylight hours are shorter, many individuals experience a change in mood and energy levels, commonly referred to as seasonal affective disorder (SAD). This condition is thought to result from a lack of exposure to natural light, which disrupts the body's production of serotonin and melatonin, leading to feelings of depression and lethargy.

The brain's recursive loops are responsible for integrating these environmental cues into the body's temporal framework, allowing individuals to adapt to changes in the external environment. For example, during the winter months, the brain may loop through sensory information about light levels and temperature, adjusting the body's internal clocks to align with the shorter daylight hours. However, when the environmental cues are insufficient, as in the case of prolonged darkness or artificial lighting, the brain's recursive loops may struggle to maintain proper synchronisation, leading to disruptions in time perception.

Similarly, individuals who live in polar regions, where daylight hours can vary dramatically between seasons, often experience significant changes in their circadian rhythms. During the polar winter, when darkness can last for months, individuals may experience sleep disturbances, mood changes, and a distorted sense of time, as the brain's recursive loops struggle to align internal clocks with the external environment.

The Recursive Mind model provides a framework for understanding how the brain integrates seasonal and environmental information into its processing of time. By recursively looping through both internal biological rhythms and external environmental cues, the brain is able to maintain a coherent sense of time, even in the face of significant changes in the environment. However, when these loops are disrupted, whether through lack of light, extreme temperatures, or seasonal shifts, individuals may experience time distortions and temporal disorientation.

Hypothesis: Can Enhancing Recursive Synchronisation Improve Temporal Alignment?

A key hypothesis that emerges from the Recursive Mind model is that enhancing the brain's ability to synchronise its recursive loops with external time cues could improve temporal alignment and reduce the negative effects of circadian misalignment. By targeting the brain's ability to integrate light exposure, sleep cycles, and environmental rhythms, it may be possible to improve time perception and cognitive function in individuals who experience time-related disruptions, such as those with jet lag, shift work disorder, or seasonal affective disorder.

This hypothesis could be tested by developing interventions that focus on circadian entrainment, the process of aligning the body's internal clocks with external time cues. For example, light therapy and cognitive-behavioural therapy for sleep could be used to enhance the brain's recursive processing of light and dark cycles, improving sleep quality and reducing the negative effects of circadian disruption. By measuring changes in brain activity, sleep patterns, and time perception before and after these interventions, researchers could explore whether enhancing recursive synchronisation leads to improvements in temporal alignment and overall well-being.

The Biological Foundations of Time Perception in the Recursive Mind

In conclusion, the body's biological rhythms, from the circadian cycle to ultradian and infradian rhythms, play a crucial role in shaping how we perceive and experience time. The brain's ability to synchronise its internal clocks with external time cues is essential for maintaining a coherent sense of time, and disruptions to these rhythms can lead to significant distortions in time perception. The Recursive Mind model provides a powerful framework for understanding how the brain integrates biological and environmental information to create a dynamic and flexible sense of time.

As we move forward, we will continue to explore how the brain's recursive nature interacts with other aspects of time perception, such as consciousness, attention, and cognitive control, further deepening our understanding of how time shapes human experience across all dimensions of life.

Chapter 15: Attention, Consciousness, and Cognitive Control: Time Perception and Complex Decision-Making

The way we perceive time is deeply influenced by the brain's ability to focus attention, sustain conscious awareness, and exercise cognitive control. Time is experienced not as a constant, but as something that shifts and fluctuates depending on how engaged we are with the present moment. When attention is focused, time seems to pass more quickly; when we are distracted or anxious, it often slows down. The processes of decision-making, conscious thought, and the ability to enter flow states also shape how we experience the passage of time.

This chapter explores how attention, consciousness, and cognitive control interact to influence time perception. We will examine how the brain's recursive loops allow it to shift attention between different tasks and time scales, creating a flexible and dynamic sense of temporal flow. Additionally, we will explore the neurological mechanisms behind complex decision-making, showing how the brain integrates past experiences, current information, and future predictions to navigate time in ways that enhance efficiency and adaptability.

Attention and Time Perception: Focused Engagement vs. Distraction

Attention is one of the primary factors that shape how we perceive time. When we are deeply focused on a task, time tends to speed up, a phenomenon often referred to as time compression. Conversely, when we are distracted or disinterested, time seems to slow down, creating the experience of time dilation. The Recursive Mind model suggests that this relationship between attention and time perception is a result of the brain's ability to loop through sensory input and cognitive processes, filtering out irrelevant information and focusing on the most important stimuli.

In states of focused attention, the brain's recursive loops become streamlined, allowing it to process information more efficiently. By continuously revisiting and refining relevant sensory input, the brain creates a seamless flow of experience where time appears to pass quickly. For example, when you are fully immersed in a book, conversation, or creative activity, your brain is looping through the task at hand, updating and integrating new information without distraction. This creates the sensation that time is passing quickly because the brain is saturated with meaningful input.

In contrast, when attention is scattered or divided, such as when multitasking or when distracted by external stimuli, the brain's recursive loops must constantly shift between different streams of information. This shifting creates a sense of discontinuity in the flow of time, where each moment feels longer because the brain is unable to maintain a focused, cohesive narrative. As a result, time may feel as though it is moving slowly, with each minute stretching out as the brain struggles to stay engaged with the present moment.

The Recursive Mind model highlights the dynamic and adaptive nature of attention, showing how the brain's ability to loop back to relevant information allows it to create a flexible sense of temporal flow. Whether time feels fast or slow depends on how efficiently the brain is able to loop through the relevant inputs and maintain focus.

Consciousness and Temporal Awareness: The Subjective Flow of Time

Consciousness plays a central role in shaping how we experience time. Our awareness of the present moment, our ability to reflect on the past, and our capacity to anticipate the future all contribute to the brain's recursive processing of time. The Recursive Mind model suggests that consciousness is not a static state but rather a continuous looping process, where the brain integrates sensory input, thoughts, and memories to create a coherent sense of temporal flow.

One of the key aspects of consciousness is its ability to shift between different time scales, allowing us to move fluidly between immediate experiences and long-term goals. For example, when we reflect on a past event, the brain's recursive loops revisit the sensory details and emotional context of that experience, integrating it into the present moment. Similarly, when we plan for the future, the brain loops through predictions and hypothetical scenarios, using information from the past and present to anticipate what will happen next.

This recursive interaction between past, present, and future is what gives rise to our subjective flow of time. Consciousness allows the brain to maintain a sense of continuity across time, even as it shifts between different temporal perspectives. Whether we are reflecting on the past, engaging with the present, or imagining the future, the brain's recursive loops ensure that these different temporal modes are integrated into a cohesive experience.

However, disruptions to consciousness, such as during sleep deprivation, intoxication, or certain neurological disorders, can lead to distortions in time perception. When the brain's recursive loops are unable to function efficiently, individuals may experience fragmented or disjointed time, where moments feel disconnected from one another. In extreme cases, such as during dissociative states or certain psychiatric conditions, individuals may lose their sense of time altogether, feeling as though they are outside the normal flow of temporal experience.

Cognitive Control and Decision-Making: Directing Time in Complex Situations

Cognitive control refers to the brain's ability to regulate its own processes, directing attention, memory, and decision-making to achieve specific goals. In the context of time perception, cognitive control plays a key role in how we navigate complex situations that require the integration of multiple time scales. The Recursive Mind model suggests that cognitive control relies on the brain's ability to loop through past experiences, present information, and future predictions, allowing it to make decisions that are grounded in both the immediate and long-term consequences.

When we engage in complex decision-making, such as planning a project, solving a problem, or navigating a social situation, the brain's recursive loops allow it to consider multiple outcomes and possibilities. By looping through past experiences, the brain can draw on lessons learned and apply them to the current situation. At the same time, the brain loops through future projections, evaluating potential consequences and adjusting its actions accordingly.

This recursive process of decision-making allows the brain to navigate time in a highly adaptive way, where both short-term and long-term goals are integrated into the cognitive process. For example, when planning a project, the brain must balance immediate tasks (such as gathering information or setting up a timeline) with long-term goals (such as completing the project on time or achieving a specific outcome). The brain's recursive loops allow it to shift between these different time scales, ensuring that the plan remains flexible and responsive to changing circumstances.

However, cognitive control can be overloaded in situations where there are too many variables or conflicting goals. When the brain's recursive loops are unable to process the information efficiently, individuals may experience decision fatigue, where time seems to slow down and the decision-making process becomes laboured and inefficient. This cognitive overload can lead to paralysis by analysis, where the brain becomes stuck in a recursive loop of indecision, unable to move forward.

The Recursive Mind model suggests that enhancing cognitive control, through mindfulness, cognitive training, or mental strategies, could improve time perception and decision-making by increasing the brain's ability to loop through relevant information efficiently. By training the brain to focus on the most important inputs and filter out distractions, individuals may be able to improve their ability to navigate time in complex situations.

Flow States: The Peak of Temporal Efficiency

One of the most profound examples of how attention, consciousness, and cognitive control interact to shape time perception is the experience of flow. Flow is a state of deep immersion and effortless concentration, where individuals become so fully engaged in an activity that they lose track of time. Time seems to fly by, and the individual experiences a sense of temporal compression, where hours can feel like minutes.

Flow occurs when the brain's recursive loops are operating at peak efficiency, allowing the individual to process information seamlessly and without distraction. In a flow state, the brain continuously loops through sensory input, cognitive processes, and motor actions, creating a self-sustaining cycle of engagement. The result is a sense of timelessness, where the individual is fully absorbed in the present moment and no longer conscious of the passage of time.

The Recursive Mind model suggests that flow represents a state of optimal recursion, where the brain's loops are finely tuned to the demands of the task at hand. By maintaining focus on the present moment and filtering out irrelevant information, the brain creates a coherent temporal experience that feels fast and effortless. This state of temporal efficiency is associated with feelings of joy, creativity, and productivity, as the brain is able to operate without the usual cognitive limitations that slow down time perception.

Flow is often experienced during activities that require a balance of challenge and skill, such as playing a musical instrument, engaging in sports, or working on a creative project. The brain's recursive loops are constantly adjusting to the demands of the task, creating a sense of dynamic feedback that keeps the individual engaged. This continuous loop of feedback and adjustment allows the brain to create a sense of flowing time, where each moment is seamlessly connected to the next.

Hypothesis: Can Training in Cognitive Control Enhance Time Perception in Complex Tasks?

A key hypothesis that emerges from the Recursive Mind model is that training in cognitive control, through techniques such as mindfulness meditation, cognitive-behavioural therapy, or cognitive training, could enhance time perception in complex tasks. By improving the brain's ability to loop through relevant information efficiently, individuals may be able to maintain focus, improve decision-making, and enter flow states more easily.

This hypothesis could be tested by measuring changes in time perception, decision-making efficiency, and flow states before and after cognitive training interventions. By examining how these interventions affect brain activity and cognitive performance, researchers could explore whether enhancing cognitive control leads to improvements in both time perception and overall productivity.

Attention, Consciousness, and Cognitive Control in the Recursive Mind

In conclusion, the brain's ability to manage attention, sustain conscious awareness, and exercise cognitive control plays a crucial role in shaping how we experience time. Through recursive processing, the brain is able to loop through past experiences, present sensory input, and future predictions, creating a dynamic and flexible sense of temporal flow. Whether time feels fast, slow, or fragmented depends on how efficiently the brain is able to loop through the relevant inputs and maintain focus on the task at hand.

As we move forward, we will continue to explore how the brain's recursive nature interacts with other aspects of time perception, such as memory, emotion, and cultural constructs, further deepening our understanding of how time shapes human experience across all domains of life.

Chapter 16: Creativity, Imagination, and Narrative Thinking: Time Perception and the Construction of Temporal Reality

Human perception of time is not merely a passive reflection of the external world but is actively shaped by creativity, imagination, and narrative thinking. These cognitive processes allow us to not only engage with the present moment but also to reshape the past, imagine the future, and weave these experiences into cohesive stories that give our lives meaning and direction. Through imagination and creativity, we can stretch, compress, and even transcend time, while narrative thinking enables us to construct temporal realities that align with our goals, values, and aspirations.

This chapter delves into how creativity, imagination, and narrative thinking interact with time perception, shaping how we experience the passage of time and understand our place in the temporal flow of life. By exploring how the brain's recursive loops generate mental simulations, fantasy worlds, and personal narratives, we can gain insight into how the brain constructs temporal meaning through creative and imaginative thought.

Imagination and Mental Time Travel: Navigating Past, Present, and Future

Imagination is one of the most powerful cognitive tools for reshaping our experience of time. It allows us to engage in mental time travel, where we revisit past events, simulate potential futures, or even create entirely new, fantastical worlds that exist outside the constraints of reality. Through imagination, we can transcend the limitations of the present moment, creating mental simulations that allow us to explore possibilities and make sense of our experiences in new ways.

The Recursive Mind model suggests that imagination relies on the brain's ability to loop through past experiences, current knowledge, and future projections, weaving them together into coherent mental simulations. When we imagine a future event, for example, the brain loops through memories of similar past events, integrates present sensory input, and projects those patterns forward into the future. This recursive interaction between past, present, and future creates a flexible and dynamic sense of time, where the boundaries between these temporal modes are blurred.

Imagination also allows us to reconstruct the past, offering new interpretations or alternative versions of events. For example, when we reflect on a difficult experience, the brain's recursive loops may generate an imaginative reinterpretation of the event, reframing it as a learning experience or an opportunity for growth. This ability to imaginatively reconstruct the past highlights the plasticity of time perception, showing that time is not a fixed entity but something that can be reshaped through creative thought.

The brain's recursive loops also enable us to create entirely new temporal realities through fantasy and storytelling. When we engage in creative writing, painting, or storytelling, the brain generates narrative worlds that unfold across time, allowing us to experience alternate versions of reality. In these creative processes, the brain loops through symbolic representations of time, such as characters' life journeys, plot progressions, and narrative arcs, creating temporal structures that mirror our real-world experience of time while offering new possibilities for exploration.

Narrative Thinking and Temporal Coherence: Weaving the Past, Present, and Future

Narrative thinking is central to how humans construct a coherent sense of self and time. We naturally create stories that connect our past experiences, present actions, and future goals, giving our lives a sense of temporal coherence and meaning. These narratives help us navigate the flow of time by allowing us to see our experiences as part of a larger personal story, where each moment is connected to the next in a meaningful sequence.

The Recursive Mind model shows that narrative thinking relies on the brain's recursive loops, which continuously revisit and update past experiences in light of new information. Each time we tell a story about ourselves, whether to others or in our own minds, the brain loops through the relevant memories, integrates present sensory input, and projects potential outcomes into the future. This recursive process allows us to maintain a sense of continuity across time, even as we encounter new experiences and challenges.

For example, when reflecting on a life-changing event, such as a career change or a personal loss, the brain's recursive loops allow us to weave that event into the larger story of our lives. We might frame the event as a pivotal turning point, a moment of growth, or a challenge that ultimately led to a positive outcome. This narrative framing helps us maintain a sense of temporal coherence, where even difficult or disruptive events are seen as part of a meaningful journey.

Narrative thinking also plays a key role in how we plan for the future. By creating personal narratives that link our present actions to our future goals, we are able to maintain a sense of purpose and direction. For example, when planning a long-term project, such as writing a book or starting a business, the brain's recursive loops generate a narrative arc that guides us through the various stages of the process, helping us stay motivated and focused over time. This ability to create forward-looking narratives is essential for goal-setting and self-regulation, as it allows us to see our future selves as extensions of our present actions.

Creativity and the Elasticity of Time: Stretching and Compressing Temporal Experience

Creativity allows us to bend and reshape our experience of time, creating moments where time seems to stretch or compress depending on our level of engagement with the creative process. When we are deeply immersed in a creative activity, whether painting, writing, or playing music, time often seems to fly by, as we become so absorbed in the task that we lose track of the clock. This experience of temporal compression is similar to the flow states discussed in previous chapters, where the brain's recursive loops become finely tuned to the demands of the task, creating a seamless sense of temporal flow.

Conversely, creativity can also make time feel as though it is expanding, particularly when we are engaged in imaginative thinking that stretches beyond the present moment. For example, when brainstorming ideas for a novel or conceptualising a new art project, the brain's recursive loops generate multiple possible futures, exploring different creative directions and imagining various outcomes. This imaginative process can create a sense of temporal elasticity, where the mind seems to stretch beyond the immediate constraints of time, exploring a vast array of possibilities.

The Recursive Mind model suggests that this ability to stretch and compress time through creativity is a result of the brain's recursive flexibility. By looping through different temporal perspectives, past, present, and future, the brain is able to generate new ideas and insights that transcend the immediate moment. This temporal elasticity allows us to engage with time in a highly creative and adaptive way, where the boundaries of time become fluid and responsive to the demands of the creative process.

Storytelling and Temporal Meaning: How We Make Sense of Time

Storytelling is one of the most powerful ways that humans make sense of time. Through stories, we create temporal structures that help us understand the flow of events, the relationships between cause and effect, and the passage of time itself. Whether we are telling a personal anecdote, reading a novel, or watching a film, storytelling provides a framework for organising our experience of time and making sense of the world around us.

The Recursive Mind model shows that storytelling relies on the brain's ability to loop through narrative structures, generating a cohesive sequence of events that unfolds across time. When we engage with a story, the brain loops through characters, plot developments, and symbolic themes, creating a mental simulation of the temporal flow of the narrative. This recursive process allows us to track the progression of events, anticipate future outcomes, and reflect on the deeper meanings embedded in the story.

Storytelling also allows us to explore alternate temporal realities, where events unfold in ways that differ from our real-world experience. For example, in science fiction or fantasy literature, time may move backwards, loop in cycles, or operate according to entirely different rules. These imaginative explorations of time allow us to question and expand our understanding of temporal reality, offering new perspectives on the relationship between time, memory, and existence.

In this way, storytelling serves as a powerful tool for temporal meaning-making, allowing us to engage with time not only as a linear progression but as a multidimensional and creative experience. Through storytelling, we are able to construct temporal realities that reflect our deepest hopes, fears, and aspirations, giving our lives a sense of coherence and purpose.

Hypothesis: Can Enhancing Narrative Thinking Improve Temporal Coherence and Emotional Well-Being?

A key hypothesis that emerges from the Recursive Mind model is that enhancing narrative thinking, through techniques such as journaling, storytelling, or creative writing, could improve temporal coherence and emotional well-being. By helping individuals create more coherent and meaningful personal narratives, it may be possible to reduce feelings of temporal disorientation and enhance the sense of purpose and direction.

This hypothesis could be tested by developing narrative therapy interventions that encourage individuals to reflect on their past experiences, imagine future possibilities, and create cohesive life stories that connect their present actions to their future goals. By measuring changes in time perception, emotional well-being, and narrative coherence before and after these interventions, researchers could explore whether enhancing narrative thinking leads to improvements in both time perception and overall mental health.

Creativity, Imagination, and Narrative Thinking in the Recursive Mind

In conclusion, creativity, imagination, and narrative thinking play a central role in shaping how we perceive and engage with time. Through these cognitive processes, the brain is able to transcend the immediate moment, explore multiple temporal perspectives, and construct meaningful stories that give our lives direction and coherence. The Recursive Mind model provides a powerful framework for understanding how the brain's recursive loops generate these creative and imaginative experiences, allowing us to engage with time in a dynamic and flexible way.

As we move forward, we will continue to explore how the brain's recursive nature interacts with other aspects of time perception, such as memory, emotion, and philosophical constructs, further deepening our understanding of how time shapes human experience across all dimensions of life.

Chapter 17: The Philosophy of Time: Metaphysical Perspectives and Recursive Processing

Time has always been one of the central puzzles in philosophy, debated by metaphysicians, scientists, and theologians for millennia. While we experience time as a linear progression from past to present to future, philosophers have long questioned whether time is an objective reality, an illusion, or a mental construct that emerges from human cognition. In this chapter, we will delve into the major philosophical theories of time, exploring how the Recursive Mind model aligns with and challenges these perspectives.

By examining how presentism, eternalism, and the block universe theory interact with the brain's recursive processing of time, we can gain deeper insights into the subjective and objective dimensions of time. Additionally, we will explore how phenomenology, the study of lived experience, helps us understand the brain's role in creating a coherent temporal framework.

Presentism: The Reality of the Present Moment

Presentism is the philosophical view that only the present moment is real. According to presentism, both the past and the future are illusions, they do not exist as actual entities but only as mental constructs or memories. The present is the only reality, and time is simply the transition of the present moment from one state to another.

From the perspective of the Recursive Mind model, presentism resonates with how the brain constructs the immediate experience of time. The brain's recursive loops are constantly processing sensory input in real-time, updating our awareness of the present moment while drawing on past memories to inform future decisions. In this sense, the brain is always anchored in the present, as it is the only moment we directly experience.

However, while presentism aligns with the brain's focus on the present, the Recursive Mind model also suggests that the brain is never entirely confined to the present. By looping through past experiences and projecting into future possibilities, the brain continuously integrates different temporal modes into a dynamic mental model of time. This recursive processing allows us to mentally navigate the past and future, even if we are always anchored in the present.

Therefore, while presentism may describe how we experience immediate reality, it does not fully account for the brain's ability to transcend the present and engage with temporal dimensions beyond the present moment. In this way, the Recursive Mind model suggests that while the present is central to our experience of time, it is not the only reality we interact with, as the brain is always cycling through multiple layers of time.

Eternalism: The Block Universe and Temporal Existence

In contrast to presentism, eternalism is the view that the past, present, and future all exist simultaneously in a block universe. In this model, time is not something that flows but rather a fourth dimension that exists alongside the three spatial dimensions. All moments in time, past, present, and future, are equally real, and time is viewed as a static block where events are simply located at different points.

The block universe theory challenges the common intuition that time passes and that the future is open and undetermined. Instead, it suggests that all events are fixed in time, with the flow of time being a psychological illusion created by our perception.

The Recursive Mind model provides an intriguing perspective on the block universe theory by showing how the brain's recursive loops allow us to experience temporal flow even within a static framework. While eternalism suggests that all moments in time are equally real, the brain's recursive loops create the illusion of temporal progression by constantly revisiting the past and anticipating the future. In this way, the brain generates a sense of dynamic flow, even if time itself is static.

For example, when we remember a past event, the brain does not simply retrieve the event as a static point in time; instead, it reconstructs the event, integrating it into the present moment and reshaping our understanding of it. Similarly, when we imagine the future, the brain projects possible outcomes by looping through future scenarios, creating the illusion that time is moving forward. In this way, the brain's recursive processing gives us the experience of time as a flow, even if eternalism suggests that all moments are already fixed in a block universe.

The Recursive Mind model thus offers a way to reconcile eternalism with our subjective experience of time. While time may be static in a block universe, the brain's recursive loops create the perception of temporal flow, allowing us to navigate time as though it is constantly unfolding.

The Arrow of Time: Entropy and Temporal Asymmetry

The arrow of time refers to the unidirectional flow of time from past to future, a phenomenon closely tied to the concept of entropy in thermodynamics. According to the second law of thermodynamics, entropy (or disorder) tends to increase over time, creating a natural progression from order to disorder. This increase in entropy is often seen as the driving force behind the arrow of time, giving time its asymmetry, where the past influences the future, but not the other way around.

From the perspective of the Recursive Mind model, the arrow of time is mirrored in the brain's processing of causality and event sequencing. The brain creates a sense of temporal asymmetry by continuously looping through cause-and-effect relationships, allowing us to understand how past events lead to present outcomes and how present actions will shape the future. This recursive processing creates the subjective experience of time moving in one direction, always from past to future, aligning with the arrow of time described by thermodynamics.

However, the brain's recursive loops also highlight the flexibility of time perception. While we experience time as moving forward, the brain is capable of revisiting past events and reinterpreting them in light of new information. This ability to rework the past suggests that while time's arrow may move in one direction at a physical level, the brain's recursive processing allows us to mentally reverse or reshape the past, adding a layer of plasticity to our experience of time.

In this way, the Recursive Mind model offers a nuanced view of the arrow of time, suggesting that while physical processes are governed by entropy and temporal asymmetry, the brain's recursive processing introduces an element of temporal flexibility that allows us to navigate time in more dynamic ways.

Phenomenology of Time: The Lived Experience of Temporal Flow

In phenomenology, time is not treated as an objective entity but as a lived experience, something that is deeply intertwined with consciousness and self-awareness. Phenomenologists such as Edmund Husserl and Martin Heidegger have argued that time is not something that exists independently of human experience but is instead a subjective phenomenon that emerges from our ongoing engagement with the world.

The Recursive Mind model aligns closely with the phenomenological view by emphasising the role of consciousness in creating the experience of time. Through recursive loops, the brain integrates past memories, present sensory input, and future projections into a continuous stream of consciousness, allowing us to experience time as something that flows and unfolds. This recursive processing creates a temporal narrative, where each moment is connected to the next in a meaningful sequence.

Phenomenologists also argue that time is experienced differently depending on our mental state and emotional context. For example, time may feel fast when we are engaged in a pleasurable activity or slow when we are bored or anxious. The Recursive Mind model provides a biological framework for understanding this subjective elasticity of time by showing how the brain's recursive loops adjust temporal processing based on emotional and cognitive states. When we are deeply engaged, the brain's recursive loops become streamlined, creating a sense of time compression. When we are distracted or disinterested, the loops become fragmented, leading to time dilation.

This phenomenological approach to time highlights the subjective nature of temporal experience, showing that time is not a static entity but something that is constantly constructed and reconstructed through the recursive processing of consciousness.

Time and Free Will: Recursive Processing and the Experience of Agency

The relationship between time and free will has long been a topic of philosophical debate. If the future is already determined, as suggested by deterministic theories of time, do we truly have the ability to make free choices that shape the future? On the other hand, if the future is open and undetermined, how do we navigate a world where outcomes are not fixed?

The Recursive Mind model offers a perspective on this debate by showing how the brain's recursive loops allow us to engage in future planning and decision-making. By looping through past experiences and simulating future possibilities, the brain creates a mental framework in which we can evaluate different options and make choices based on anticipated outcomes. This recursive process gives rise to the experience of agency, where we feel as though we are actively shaping the future through our decisions.

While the future may already exist in a block universe model, the brain's recursive loops allow us to experience the act of choosing among different possibilities. This suggests that free will may be an emergent property of the brain's recursive processing, giving rise to the subjective experience of agency even within a deterministic framework. In this way, the Recursive Mind model provides a potential bridge between philosophical determinism and the experience of free will. Although time may be laid out in its entirety in a block universe, the brain's ability to recursively simulate potential outcomes and evaluate choices allows us to engage with time in a manner that feels active and dynamic.

This recursive ability to navigate multiple possibilities, evaluate them, and take action creates a felt sense of free will. Even if the future is ultimately part of a pre-existing framework, the brain's recursive loops make the process of choosing feel like an exploration of new, uncharted territory. It allows for a subjective experience where the individual is continually crafting their own path through time, with a sense of self-direction and control.

Time, Causality, and Recursive Mental Models

Causality, understanding how one event leads to another, is central to how we organise and navigate time. The human brain is remarkably attuned to recognising cause-and-effect relationships, allowing us to predict outcomes and make sense of the sequence of events around us. Philosophers have debated the nature of causality for centuries, questioning whether causality is an inherent feature of reality or something we impose on our experiences to make sense of them.

The Recursive Mind model suggests that our experience of causality is shaped by the brain's ability to loop through past experiences and use them to inform current decisions and future predictions. By recursively revisiting cause-and-effect patterns from the past, the brain builds a mental model of how the world works, which it then applies to present and future situations.

For example, when we experience a particular event, such as a glass falling from a table and shattering, the brain loops through the visual input and past memories of similar events, recognising the pattern of causality: the glass fell, hit the ground, and broke. This recursive mental model is then reinforced by the brain's prediction that similar events will happen again under similar circumstances. As a result, our sense of causal structure in the world is built on the brain's recursive processing of time, where past experiences are integrated into present and future reasoning.

Moreover, when we consider the possibility of alternative outcomes, for example, imagining that we could have caught the glass before it broke, the brain uses recursive loops to simulate different causal paths, exploring what might have happened if the sequence of events had been different. This ability to mentally rework causal chains allows us to reflect on choices and consequences, deepening our understanding of how events unfold in time.

In this way, the brain's recursive processing not only supports our experience of time but also our ability to understand and engage with causal dynamics. The brain loops through causal models of the world, revising and refining them based on ongoing experiences, creating a flexible and adaptive framework for navigating time and causality.

Hypothesis: Can Recursive Processing Illuminate the Philosophy of Time?

A key hypothesis that emerges from the Recursive Mind model is that recursive processing in the brain could offer new insights into longstanding philosophical questions about time, causality, and free will. By studying how the brain processes temporal information and cause-and-effect relationships, we may be able to better understand how our subjective experience of time aligns with, or diverges from, philosophical theories.

This hypothesis could be explored through neuroscientific research that examines the brain's activity during tasks that involve memory recall, future planning, and causal reasoning. By investigating how the brain's recursive loops process temporal information in these contexts, researchers could explore whether the brain's cognitive architecture aligns more closely with presentism, eternalism, or the block universe theory.

Additionally, by studying individuals who experience temporal distortions, such as those with neurological conditions that affect time perception, researchers could gain further insights into how the brain constructs the flow of time and how it processes causality in different mental states.

Philosophical Time and the Recursive Mind

In conclusion, the Recursive Mind model provides a powerful framework for exploring how the brain processes time, causality, and free will, all central concerns in philosophical debates about the nature of time. By examining how the brain's recursive loops allow us to navigate past, present, and future experiences, we can gain new insights into whether time is an absolute reality, a mental construct, or something in between.

The recursive nature of the brain allows us to simulate possibilities, revisit the past, and anticipate the future, creating a dynamic and adaptive temporal experience. This suggests that while philosophical theories may describe time in abstract terms, the human brain plays an active role in shaping how time is perceived and experienced in everyday life.

As we move forward, we will continue to explore how the Recursive Mind model intersects with both scientific and philosophical perspectives on time, deepening our understanding of how the brain creates meaning from the passage of time and how this process informs our understanding of reality.

Chapter 18: Time and Memory: Constructing and Reconstructing Temporal Narratives

Memory is the key to how we perceive and understand the passage of time. Through memory, the brain creates a narrative that links the past to the present, allowing us to construct a coherent sense of self and to make sense of our place in the world. However, memory is not a perfect, static record of past events. It is a dynamic process through which the brain continuously reconstructs and reinterprets our experiences, shaped by present concerns, emotions, and expectations for the future.

In this chapter, we will explore how memory interacts with time perception, focusing on the brain's recursive processing of memories, and how this contributes to the construction of temporal narratives. We will examine the different types of memory, such as episodic, semantic, and procedural memory, and how they contribute to our understanding of the flow of time. Additionally, we will investigate the concept of false memories and memory distortions, which highlight the flexible, reconstructive nature of memory and its role in shaping our perception of time.

Episodic Memory: The Mental Time Machine

Episodic memory refers to the brain's ability to recall specific events from our personal past, often with rich details of the who, what, when, and where of the experience. It allows us to mentally travel back in time, revisiting the sights, sounds, and emotions of previous moments in our lives. Unlike other forms of memory, such as semantic memory (general knowledge) or procedural memory (skills and habits), episodic memory is intimately tied to the subjective experience of time.

The Recursive Mind model suggests that episodic memory operates through recursive loops that allow the brain to revisit and update past experiences in light of new information. When we recall a memory, the brain loops through the original event, integrating it with the present context, thoughts, and emotions. This process of recursive reconstruction is not passive; each time we remember an event, it is slightly modified by our current state, feelings, and understanding. This explains why memories often change subtly over time, they are not fixed snapshots, but dynamic reconstructions.

For example, consider a memory of a childhood birthday party. Initially, the memory might be vivid, full of details about the setting, people, and emotions. However, as time passes and life experiences accumulate, the memory may shift. New insights, such as a deeper understanding of relationships or personal growth, can influence how that event is remembered. What was once simply a happy day may, in hindsight, take on additional meanings, perhaps as a formative moment or a marker of familial dynamics that were not clear at the time. The brain's recursive loops continuously reshape this memory, integrating new layers of meaning each time it is recalled.

Episodic memory, therefore, functions as a mental time machine, allowing us to revisit the past and project elements of it into the present. This process supports the creation of a temporal narrative, where our past informs our current identity and influences how we imagine the future. Episodic memory is crucial for maintaining a sense of continuity across time, grounding our experience in a coherent personal story.

Semantic and Procedural Memory: Beyond the Self

While episodic memory is tied to personal experiences, semantic memory encompasses our general knowledge about the world, facts, concepts, and meanings that are not tied to specific personal events. For example, knowing that Paris is the capital of France, or that gravity causes objects to fall, are pieces of semantic memory. Semantic memory is less about mentally traveling through time and more about static knowledge that we can access at any moment.

Semantic memory contributes to our perception of time by providing a contextual framework for events. For instance, when we recall an episodic memory, such as a family vacation, our understanding of historical, geographical, and social facts, drawn from semantic memory, helps us place that experience in a broader temporal framework. This allows us to situate personal events within the larger context of time, making sense of how personal histories intersect with broader cultural or historical timelines.

Procedural memory, on the other hand, involves the unconscious memory of how to perform tasks, such as riding a bike, playing the piano, or typing on a keyboard. These memories are deeply ingrained and typically do not require conscious recall. While procedural memory operates largely outside of our conscious awareness, it still interacts with time perception in subtle ways. Mastering a skill often involves repetition over time, and procedural memory tracks these repetitions, allowing for the refinement of skills over long periods. In a sense, procedural memory represents time through progress and learning, tracking how repeated actions lead to smoother and more efficient performance.

Memory Distortions: The Flexibility of Reconstructive Memory

One of the most fascinating aspects of memory is its flexibility, particularly in how it can be influenced and reshaped by new information, emotions, or social cues. Memory distortions, such as false memories, highlight how reconstructive memory can be prone to error. False memories occur when the brain mistakenly recalls an event that either did not happen or happened differently than remembered. These distortions can result from a variety of factors, including suggestion, imagination, or misleading information from others.

The Recursive Mind model explains false memories as a byproduct of the brain's recursive loops, which continuously reprocess and reinterpret past experiences. Each time we recall a memory, the brain reconstructs it based on current input, including emotions, beliefs, and new knowledge. If the brain receives inaccurate information or if it tries to fill in gaps in memory, it may introduce distortions that alter the memory. Over time, these alterations can become integrated into the narrative of that memory, making it difficult to distinguish between fact and fiction.

For example, if someone tells you a story about an event from your childhood, and it conflicts with your own memory, your brain may eventually incorporate elements of that story into your own recollection. The brain's recursive loops will process the new input and modify the original memory to fit the new narrative. This reconstructive process is why eyewitness testimonies are notoriously unreliable, people's memories of events can be easily influenced by external suggestions or personal biases.

Memory, Time, and the Self: Constructing Temporal Identity

At the heart of memory's interaction with time is the role it plays in constructing our temporal identity. Our memories are not just a record of the past; they form the foundation of our sense of self. Through memory, we connect who we are now with who we were in the past, creating a continuous thread that stretches across time. This temporal identity is essential for maintaining a coherent narrative of our lives and for understanding how we have changed and developed over time.

The Recursive Mind model suggests that memory is the primary tool the brain uses to create temporal coherence. Through recursive loops, the brain revisits and updates past experiences, integrating them into the present and projecting them into the future. This recursive process allows us to create a personal narrative that connects our past actions, present circumstances, and future goals. In essence, memory helps us make sense of the flow of time by giving us a structured framework in which our experiences and decisions are organized into a meaningful whole.

For example, when we think about our future goals, we often draw on our past experiences to shape our expectations and aspirations. We may reflect on what we have learned from previous successes or failures, using those insights to inform our choices moving forward. In this way, memory not only helps us recall the past but also shapes our perception of the future, creating a bridge between different moments in time.

Hypothesis: Can Recursive Memory Processing Improve Temporal Awareness?

A key hypothesis that emerges from the Recursive Mind model is that improving the brain's ability to process memory recursively could enhance temporal awareness and reduce memory distortions. By strengthening the connections between episodic, semantic, and procedural memory, it may be possible to improve the brain's ability to track time, maintain temporal coherence, and avoid false memories.

This hypothesis could be tested by developing cognitive training programs that focus on enhancing the brain's recursive memory circuits. Such interventions could involve exercises designed to improve memory recall, memory updating, and the ability to distinguish between accurate and distorted memories. By enhancing recursive memory processing, individuals may develop a clearer sense of temporal flow and a more accurate understanding of how their personal narrative unfolds across time.

Memory as a Time Machine in the Recursive Mind

In conclusion, memory is not just a passive record of the past; it is an active, reconstructive process that allows the brain to navigate time. Through recursive loops, the brain continuously revisits and updates past experiences, integrating them into the present and using them to anticipate the future. This recursive processing allows us to create coherent temporal narratives, linking our past, present, and future into a unified sense of self.

As we move forward, we will continue to explore how the brain's recursive nature interacts with other aspects of time perception, such as emotion, attention, and cultural influences, further deepening our understanding of how the brain creates meaning from the passage of time.

Chapter 19: Emotion and Time: The Recursive Loops of Emotional Memory and Time Perception

Emotion plays a profound role in shaping our perception of time. Whether time feels like it's moving too quickly or too slowly often depends on our emotional state. Moments of joy seem to fly by, while painful experiences can make time feel like it's dragging. These variations in time perception are not merely the result of external events but are deeply connected to how the brain processes emotions in relation to temporal awareness.

In this chapter, we will explore the interaction between emotion and time perception, examining how emotions influence the brain's recursive processing of time. We will investigate how emotional memory affects the way we perceive past events, how emotional intensity can distort our sense of temporal flow, and how certain emotional states, such as anxiety, grief, or happiness, affect our experience of time. Additionally, we will delve into the neurological mechanisms that link emotion with time perception, focusing on brain regions such as the amygdala and prefrontal cortex.

The Influence of Emotion on Time Perception

One of the most significant ways emotion shapes our perception of time is through its ability to expand or compress our sense of temporal flow. During moments of intense emotion, such as excitement, fear, or sorrow, time often feels distorted. For example, people commonly report that time seems to slow down during a car accident or other life-threatening situations, while positive experiences, like spending time with loved ones, make time feel as though it has sped up.

The Recursive Mind model suggests that these distortions occur because the brain's recursive loops process emotional experiences differently from neutral ones. When we experience strong emotions, the brain's amygdala, the region responsible for emotional regulation and responses, activates, heightening the brain's focus on sensory input and cognitive processing. The recursive loops continuously revisit and amplify the emotional significance of the moment, leading to a distortion in time perception.

For example, during a moment of fear, the brain enters a recursive loop where it repeatedly analyses the environment for potential threats, enhancing vigilance and prolonging the perception of time. This heightened state of alertness creates a subjective experience of time slowing down, allowing the individual to process more information in a short period.

On the other hand, during moments of joy or engagement, the brain's reward circuits, involving regions such as the nucleus accumbens and dopamine pathways, create recursive loops that reinforce the positive emotional experience, leading to a sense of time compression. In these moments, the brain filters out distractions and focuses on the enjoyable activity, making time feel like it is passing quickly. These recursive processes allow the brain to prioritize emotional significance, shaping how we experience the passage of time.

Emotional Memory and the Reconstruction of the Past

Emotional memory is particularly powerful in shaping how we remember past events and how we interpret those memories in relation to time. Emotionally charged events, whether positive or negative, tend to be remembered more vividly than neutral events, and they often feel closer in time, even if they occurred many years ago. This phenomenon occurs because the brain's recursive loops continuously revisit these emotionally significant memories, reinforcing and reinterpreting them each time they are recalled.

For example, a person who experienced a traumatic event may feel as though the memory is fresh and immediate, even if the event occurred long ago. The brain's recursive loops continuously replay the emotional intensity of the trauma, making the memory feel present rather than confined to the past. This is particularly evident in conditions like post-traumatic stress disorder (PTSD), where traumatic memories intrude into daily life, distorting the individual's sense of time and causing past events to bleed into the present.

On the other hand, positive emotional memories, such as a wedding day, a vacation, or a major achievement, are often remembered as timeless or fleeting. These memories may feel as though they passed too quickly, even if they were long in duration. The brain's recursive processing of these positive emotional memories tends to compress time, making the memory feel like a brief but vivid moment in the individual's temporal narrative.

This flexibility in emotional memory is a result of the brain's ability to continuously reconstruct the past through recursive loops. Each time an emotional memory is recalled, the brain revisits the emotional context of the event, modifying the memory based on current thoughts, feelings, and experiences. This recursive reconstruction allows the brain to shape the temporal narrative of our lives, integrating emotionally significant events into a cohesive story that influences how we perceive time.

The Role of the Amygdala and Prefrontal Cortex in Emotional Time Perception

The brain regions most closely associated with the interaction between emotion and time perception are the amygdala and the prefrontal cortex. The amygdala is primarily responsible for processing emotions, especially fear, anxiety, and reward, while the prefrontal cortex is involved in executive function, decision-making, and attention regulation.

During emotional experiences, the amygdala becomes highly active, triggering the brain's fight-or-flight response in moments of fear or danger. The amygdala sends signals to the prefrontal cortex, which helps modulate the brain's focus on the emotional event. This interaction between the amygdala and the prefrontal cortex creates recursive loops that influence how we process and remember emotional experiences.

In the context of time perception, the amygdala's role is to heighten the brain's awareness of emotionally significant moments, amplifying the perception of time during intense emotions. The prefrontal cortex, on the other hand, helps regulate and balance this emotional input, ensuring that we maintain a coherent sense of time even during emotionally charged moments. However, when the prefrontal cortex is unable to fully regulate the amygdala's activity, such as during extreme stress or trauma, our sense of time can become distorted, leading to the experience of time dragging or accelerating.

Emotion, Memory, and the Sense of Self

Emotion is not only essential for shaping our perception of time in the moment but also plays a central role in how we construct a temporal narrative of our lives. Emotional memories form the foundation of our sense of self, linking key events from our past to our present identity. The brain's recursive loops allow us to revisit these emotional memories, updating them in light of new experiences and incorporating them into the ongoing story of who we are.

For example, a person reflecting on their personal growth may loop through emotionally significant memories, such as overcoming a challenge or achieving a goal, integrating these memories into a larger narrative of success and resilience. This recursive processing of emotional memories helps us make sense of the passage of time, allowing us to create a coherent narrative that connects the past with the present and future.

However, when emotions such as grief, guilt, or regret dominate our emotional memory, they can lead to a distorted sense of time, where the individual feels stuck in the past, unable to move forward. In these cases, the brain's recursive loops continuously revisit the painful memory, reinforcing the emotional weight of the experience and making it difficult to engage with the present or imagine a positive future. This is why individuals suffering from depression or trauma often report feeling as though they are trapped in a loop of negative emotions, with time seeming to stand still.

Hypothesis: Can Emotional Regulation Improve Temporal Perception?

A key hypothesis that emerges from the Recursive Mind model is that improving the brain's ability to regulate emotion, particularly through interventions such as mindfulness meditation or cognitive-behavioural therapy (CBT), could enhance time perception. By helping individuals manage their emotional responses and focus attention on the present moment, these interventions may reduce the temporal distortions caused by intense emotional states.

This hypothesis could be tested by measuring changes in time perception, emotional regulation, and memory accuracy before and after emotional regulation interventions. By examining how these interventions affect brain activity, particularly in the amygdala and prefrontal cortex, researchers could explore whether improving emotional regulation leads to a more balanced and accurate perception of time.

Emotion and Time in the Recursive Mind

In conclusion, emotion plays a crucial role in shaping how we perceive and engage with time. Through recursive loops, the brain continuously revisits emotionally significant events, distorting or amplifying our sense of time depending on the emotional intensity of the moment. Whether time feels fast, slow, or timeless, our emotions influence how we navigate the flow of time, and the recursive nature of emotional processing allows the brain to create a dynamic and adaptive experience of time.

As we move forward, we will continue to explore how the brain's recursive nature interacts with other aspects of time perception, such as attention, consciousness, and cultural influences, further deepening our understanding of how time shapes human experience at both the individual and collective levels.

Chapter 20: Cultural Frameworks and Time: The Influence of Society on Temporal Perception

Time is not only a biological and psychological phenomenon but also a cultural construct. Across different societies, time is understood, valued, and organised in diverse ways, leading to variations in how individuals experience and perceive its passage. From the fast-paced, clock-driven world of Western industrialised societies to the more fluid and flexible sense of time found in many indigenous cultures, the perception of time is shaped by social norms, historical contexts, and cultural frameworks.

In this chapter, we will explore how cultural models of time influence the brain's recursive processing of temporal information. We will examine how different societies conceptualise time as linear or cyclical, how the use of clocks and calendars imposes external time structures on individuals, and how modern technology and globalisation have further complicated our relationship with time. The Recursive Mind model provides a lens for understanding how the brain adapts to these external time structures, integrating cultural time frameworks into individual time perception.

Linear vs. Cyclical Time: Contrasting Cultural Perspectives

One of the most fundamental distinctions in cultural models of time is the difference between linear and cyclical perceptions of time. In many Western societies, time is viewed as linear, a one-way arrow moving from the past to the present and into the future. This perspective underpins much of the Western emphasis on progress, planning, and the importance of efficiency in both personal and societal development. In this model, time is seen as a limited resource that must be used wisely, often measured in terms of productivity and achievement.

In contrast, many Eastern and indigenous cultures view time as cyclical, where life, nature, and events occur in repeating patterns or cycles. This cyclical understanding of time is rooted in natural phenomena, such as the seasons, moon phases, or the cycle of life and death, and leads to a more flexible and adaptive relationship with time. In this model, time is not necessarily a resource to be controlled or optimised but rather something to be lived within, allowing for a focus on balance, harmony, and rhythmic flow.

The Recursive Mind model highlights how the brain can adapt to these cultural frameworks, using recursive loops to either process time as linear or cyclical, depending on cultural influences. In a linear time model, the brain's recursive loops may focus on sequencing events in a forward-moving direction, reinforcing the idea that the past is behind us and the future lies ahead. The brain's recursive processing helps individuals organise their actions and decisions based on anticipated future outcomes, prioritising goal-setting and efficiency.

In cyclical time cultures, the brain's recursive loops may focus on recognising patterns and repeating cycles, allowing individuals to view time as something that returns rather than progresses in a straight line. This mode of processing encourages individuals to see the interconnectedness of events, with past experiences often revisiting the present in new forms. The brain adapts by focusing on balance and adaptability rather than linear progress, creating a sense of timelessness or continuity across cycles.

Monochronic vs. Polychronic Cultures: How Societies Structure Time

Another key cultural distinction in time perception is between monochronic and polychronic cultures. In monochronic cultures, such as those found in the United States, Germany, and much of Northern Europe, time is seen as discrete and sequential, with a strong emphasis on schedules, punctuality, and doing one thing at a time. Time is organised into units, and people are expected to adhere strictly to timetables and deadlines. This approach to time reflects a high degree of order and control, where time is treated as a commodity that can be saved, spent, or wasted.

In contrast, polychronic cultures, such as those found in Latin America, the Middle East, and many African countries, approach time with much more flexibility. In polychronic cultures, it is common to engage in multiple activities simultaneously, and the focus is often on relationships and social interactions rather than rigid adherence to schedules. Time is seen as fluid, and people are expected to be adaptable to changing circumstances.

The Recursive Mind model suggests that the brain's recursive loops adjust to these different cultural norms. In monochronic cultures, the brain may become highly focused on task-oriented processing, where recursive loops track the progress of individual tasks in a sequential manner, ensuring that time is managed efficiently. The brain is trained to loop back through deadlines and time frames, reinforcing the cultural expectation that time must be structured and optimised.

In polychronic cultures, the brain's recursive loops may function in a more holistic and multidimensional way, allowing individuals to switch between different tasks and social interactions fluidly. The brain loops through multiple streams of information, creating a sense of temporal fluidity, where tasks and events overlap and time is experienced as abundant rather than scarce. This flexible approach allows individuals to prioritise social connections and contextual factors over rigid scheduling.

The Impact of Technology and Globalisation on Time Perception

In the modern world, technology and globalisation have significantly transformed how societies perceive and organise time. The invention of mechanical clocks, digital devices, and global communication networks has created a world where time is increasingly standardised and regulated. With the rise of the internet, smartphones, and social media, people are now expected to respond to information and tasks in real-time, accelerating the pace of life and creating a sense of time scarcity.

The Recursive Mind model offers insights into how the brain adapts to the compression of time brought about by modern technology. The brain's recursive loops are continuously interrupted by incoming stimuli, emails, notifications, and social media updates, which require constant re-evaluation and prioritisation. This creates a state of cognitive overload, where the brain's recursive loops are strained by the need to process multiple streams of information simultaneously.

As a result, time may feel as though it is moving faster, with individuals struggling to keep up with the demands of a 24/7 digital world. This acceleration of time can lead to feelings of stress, anxiety, and burnout, as the brain's recursive loops are constantly forced to manage the flow of new information. However, the brain's adaptive nature also allows for the development of strategies to manage this accelerated pace, such as digital detoxing, mindfulness practices, or time management techniques that prioritise focus and attention.

Globalisation has further complicated the experience of time by creating a world where people from different cultures and time zones must interact and collaborate. The brain's recursive loops must adjust to multiple temporal frameworks, navigating between local and global time, often requiring individuals to work across different schedules, negotiate time zone differences, and synchronise activities with people in other parts of the world.

The Recursive Mind model suggests that this global synchronisation places new demands on the brain's ability to loop through multiple cultural models of time. For individuals working in global environments, the brain's recursive loops may need to shift between monochronic and polychronic modes of time management, depending on the cultural context and the specific demands of the situation.

Cultural Constructs of Time and the Sense of Self

Cultural models of time do not only influence how we organise our daily activities; they also shape our sense of self and our relationship with time as individuals. In societies that emphasise linear time, personal identity is often tied to a narrative of progress and achievement. Individuals are encouraged to view their lives as a series of goals and milestones, with success measured by how efficiently time is used and how much is accomplished.

In contrast, in cultures that emphasise cyclical time, personal identity may be more closely tied to the rhythms of nature and community. Individuals may be encouraged to see their lives as part of a larger cycle, where personal experiences are woven into the broader patterns of life, death, and renewal. Time is seen as something that flows naturally, and personal growth may be understood in terms of balance and harmony rather than linear progression.

The Recursive Mind model suggests that the brain's recursive loops play a key role in shaping how individuals construct their temporal identity based on cultural norms. In linear time cultures, the brain's recursive loops may focus on goal-setting and future planning, reinforcing the idea that the individual's identity is built through forward movement. In cyclical time cultures, the brain may loop through repeating patterns of experience, creating a sense of continuity with the past and future.

Hypothesis: Can Cultural Adaptation Improve Time Perception and Well-Being?

A key hypothesis that emerges from the Recursive Mind model is that cultural adaptation, the ability to shift between different cultural models of time, could enhance time perception and overall well-being. By developing the ability to navigate both monochronic and polychronic time frameworks, or to balance linear and cyclical perspectives, individuals may be better equipped to manage the demands of a rapidly changing, globalised world.

This hypothesis could be tested by examining how individuals from different cultural backgrounds perceive time and how their time management skills and overall well-being are affected by cultural adaptation interventions. Such interventions might include cross-cultural training or mindfulness practices that help individuals switch between different time perspectives depending on the context.

Culture and Time in the Recursive Mind

In conclusion, culture plays a profound role in shaping how we perceive, value, and organise time. Through recursive loops, the brain adapts to different cultural models of time, whether linear or cyclical, monochronic or polychronic. These cultural frameworks influence not only how we experience time in daily life but also how we construct our sense of self and navigate our place in the world.

The Recursive Mind model provides a powerful framework for understanding how the brain processes cultural time frameworks, offering insights into how individuals adapt to different cultural norms and how globalisation and technology have transformed our relationship with time. As we move forward, we will continue to explore how the brain's recursive nature interacts with other dimensions of time perception, further deepening our understanding of how time shapes human experience at both individual and societal levels.

Chapter 21: Social Interaction and Time: The Collective Experience of Temporal Awareness

While time is often perceived as a personal, subjective experience, it also plays a crucial role in social interactions and group dynamics. Our sense of time is shaped by the social contexts in which we live, work, and interact with others. From shared schedules and societal rhythms to collective time markers such as holidays and festivals, our perception of time is intertwined with the expectations and norms of the people around us.

In this chapter, we will explore how social frameworks influence our perception of time and how the brain's recursive loops process these collective experiences. We will investigate how social interactions can alter the speed at which we perceive time, how groups synchronize their actions and sense of time, and how cultural rituals and social institutions create shared temporal realities. The Recursive Mind model will be applied to understand how these social influences interact with individual cognition to produce a collective temporal awareness.

The Social Synchronization of Time: Collective Rhythms and Temporal Coordination

In any society, time must be synchronized to some degree to allow people to coordinate their activities and interact efficiently. This synchronization of time is seen in the use of work schedules, school calendars, and social appointments, which allow individuals to align their actions with others in a shared temporal framework. The need for social coordination often drives the standardization of time, creating collective rhythms that regulate daily life.

The Recursive Mind model suggests that the brain's recursive loops are deeply influenced by these collective rhythms. By looping through social cues, such as the actions and expectations of others, the brain adjusts its own internal clock to align with the external social environment. This social synchronization allows individuals to navigate group activities, coordinate with others, and participate in the collective experience of time.

For example, when people work in an office or participate in a group project, their sense of time becomes attuned to the group's schedule and shared goals. The brain's recursive loops continuously monitor social interactions and adjust the individual's perception of time based on the pace and flow of the group's activities. This leads to a phenomenon known as temporal entrainment, where individuals unconsciously align their actions and sense of time with the collective rhythm of the group.

Temporal entrainment is especially evident in activities that require synchronized timing, such as playing in a musical ensemble, dancing in unison, or participating in a sporting event. In these contexts, the brain's recursive loops enable individuals to anticipate the actions of others and adjust their own timing accordingly, creating a seamless experience of collective temporal flow. This synchronization allows the group to function as a cohesive unit, with each member contributing to the overall rhythm of the activity.

Social Influence on Time Perception: How Group Dynamics Affect Temporal Flow

Social interactions can also alter an individual's perception of time. Research has shown that time perception is often influenced by the presence of others, with time seeming to pass more quickly or slowly depending on the nature of the social interaction. For example, when people are engaged in a stimulating conversation or enjoying a social gathering, time may feel as though it is passing quickly. On the other hand, in uncomfortable or awkward social situations, time may seem to drag.

The Recursive Mind model explains these changes in time perception through the brain's recursive loops, which continuously process social and environmental cues. In positive social interactions, the brain's recursive loops are highly engaged, looping through the emotional and cognitive stimuli of the conversation or activity. This intense focus on the present moment compresses the perception of time, making it feel as though time is passing quickly. The brain's recursive processing reinforces the enjoyment of the social experience, leading to a sense of time acceleration.

Conversely, in negative or stressful social situations, the brain's recursive loops may focus on discomfort or anxiety, heightening awareness of each passing moment. This increased attention to time can lead to time dilation, where individuals feel as though time is moving slowly. The brain loops through negative emotions and social stressors, reinforcing the perception that the experience is dragging on.

Group dynamics also play a role in how time is perceived in social settings. In groups with a strong sense of cohesion and shared purpose, time may pass more quickly as individuals become absorbed in the collective activity. In contrast, in groups marked by conflict or disagreement, time may feel slower as individuals become more aware of social tensions and the passing of time.

Shared Temporal Markers: Rituals, Festivals, and Collective Time

Cultural rituals, festivals, and social institutions create shared temporal markers that shape the collective experience of time. These events provide a structure for the passage of time, allowing individuals to align their personal timelines with broader societal rhythms. Holidays, religious ceremonies, and national observances create a sense of communal time, where the entire community participates in a shared temporal experience.

The Recursive Mind model suggests that these shared temporal markers influence the brain's recursive loops by providing regular, predictable patterns of time. When individuals participate in a cultural ritual or celebration, their brain loops through the symbolic meaning of the event, reinforcing their connection to the community and the collective temporal framework. This recursive processing helps individuals orient themselves within the broader social structure, allowing them to experience time not only as an individual phenomenon but as part of a larger, communal narrative.

For example, religious festivals such as Christmas, Diwali, or Ramadan mark specific points in the annual calendar, providing a collective rhythm that shapes the experience of time for those who participate. These events create a sense of anticipation and reflection, as the brain loops through the rituals and symbols associated with the festival. This recursive processing reinforces the individual's connection to the community, helping them situate their personal experiences within the larger cultural timeline.

Similarly, national holidays such as Independence Day or Remembrance Day serve as shared temporal markers that link individual experiences to the historical memory of the nation. The brain's recursive loops allow individuals to revisit the past, whether through commemorative events or historical narratives, while also projecting the meaning of the event into the future. This recursive interaction between past, present, and future helps create a sense of temporal continuity within the collective experience of time.

Social Acceleration: The Impact of Modern Life on Collective Time

In the modern world, social time has been significantly influenced by the forces of industrialisation, globalisation, and technology. These forces have led to what some scholars call social acceleration, the speeding up of social life, where individuals and groups are expected to do more in less time. This phenomenon is most apparent in urban, industrialised societies, where the pace of life is governed by work schedules, deadlines, and the constant flow of digital communication.

The Recursive Mind model suggests that the brain's recursive loops are deeply affected by this social acceleration, as the need to process multiple streams of information and manage competing demands strains the brain's cognitive resources. Social acceleration leads to a sense of time scarcity, where individuals feel that there is never enough time to accomplish everything. The brain loops through tasks, appointments, and obligations, creating a feeling of temporal pressure that can lead to stress and anxiety.

However, the brain's recursive loops also allow for adaptation to this accelerated pace. Through cognitive strategies such as prioritisation, time management, and mindfulness, individuals can learn to manage the demands of social acceleration more effectively. By focusing the brain's recursive loops on the most important tasks and filtering out unnecessary distractions, individuals can create a sense of temporal control, even in the face of increasing social pressures.

Hypothesis: Can Strengthening Social Synchronization Improve Well-Being and Time Perception?

A key hypothesis that emerges from the Recursive Mind model is that improving an individual's ability to synchronize with collective social rhythms could enhance well-being and improve time perception. By strengthening the brain's ability to align with group dynamics, shared temporal markers, and social schedules, individuals may experience greater temporal coherence and social connectedness.

This hypothesis could be tested by examining how individuals who participate in collective activities, such as group sports, music ensembles, or community rituals, experience time compared to those who primarily engage in individual activities. Researchers could explore whether participation in socially synchronized activities leads to improvements in time perception, social cohesion, and overall mental health.

Conclusion: Social Interaction and Time in the Recursive Mind

In conclusion, social interactions and group dynamics play a central role in shaping how we perceive and experience time. Through recursive loops, the brain adjusts to the rhythms and expectations of the social environment, allowing individuals to synchronize their actions and sense of time with the collective. Whether through temporal entrainment in group activities, the influence of cultural rituals, or the pressures of social acceleration, the brain's recursive processing helps create a shared temporal reality that shapes both individual and collective experiences of time.

As we move forward, we will continue to explore how the brain's recursive nature interacts with other aspects of time perception, such as memory, emotion, and philosophy, further deepening our understanding of how time shapes human experience at both individual and social levels.

Chapter 22: Technology, Artificial Intelligence, and the Digital Age: Transforming Our Experience of Time

In the modern world, technology, particularly digital communication and artificial intelligence, has dramatically transformed how we experience and perceive time. The pace of life has accelerated, information is processed at unprecedented speeds, and the boundaries between work and leisure, day and night, have become increasingly blurred. As we navigate an era where we are constantly connected to digital devices and surrounded by instantaneous information, our perception of time has evolved in ways that challenge traditional models of temporal awareness.

In this chapter, we will explore how technology and artificial intelligence influence the brain's recursive processing of time. We will investigate how the internet, smartphones, and social media contribute to a sense of time compression, how artificial intelligence is reshaping our relationship with time through automation and predictive algorithms, and how the digital age has led to new forms of temporal dissonance. The Recursive Mind model will help explain how the brain adapts to these changes, integrating the accelerated pace of modern life with our cognitive architecture.

The Acceleration of Time in the Digital Age: Time Compression and Cognitive Overload

One of the most profound effects of the digital age is the sense that time is moving faster than ever before. With the advent of high-speed internet, instant messaging, and on-demand entertainment, we have grown accustomed to receiving information and completing tasks at a rate that would have been unimaginable just a few decades ago. This acceleration of time is particularly evident in our work environments, where the expectation for immediate responses and rapid turnaround has created a constant sense of urgency.

The Recursive Mind model suggests that this time compression is the result of the brain's recursive loops being overloaded by the sheer volume of digital information. In the pre-digital world, the brain's recursive loops processed information at a slower, more manageable pace, allowing for reflection and deliberation. However, in the digital age, the brain is bombarded with a continuous stream of emails, notifications, social media updates, and other forms of digital input. As a result, the brain's recursive loops must work faster to process, categorise, and respond to this influx of information.

This rapid processing of information creates a feeling of cognitive overload, where individuals struggle to keep up with the demands of the digital world. The brain's recursive loops become strained, leading to feelings of stress and anxiety as individuals attempt to navigate multiple streams of information simultaneously. The experience of time compression is a byproduct of this cognitive strain, where the brain is so focused on processing information that it creates the sensation that time is passing quickly, even when the external world remains unchanged.

The Impact of Artificial Intelligence on Time Perception

As artificial intelligence (AI) continues to develop, its impact on time perception becomes increasingly significant. AI systems, with their ability to process massive amounts of data, automate complex tasks, and predict future outcomes, are reshaping the way we interact with time. From AI-driven personal assistants that schedule meetings and remind us of deadlines to machine learning algorithms that optimise business processes, AI is changing how we manage and experience time.

The Recursive Mind model provides insight into how AI interacts with the brain's recursive loops to influence time perception. By automating many of the tasks that once required conscious attention, AI frees up the brain's recursive loops, allowing individuals to focus on more creative or strategic activities. For example, AI-powered tools can handle data analysis, routine administrative tasks, or even predictive maintenance, allowing individuals to redirect their cognitive resources toward more value-driven activities.

However, the increased reliance on AI also creates a paradoxical relationship with time. On the one hand, AI allows us to accomplish more in less time, creating the sensation of time expansion, where individuals feel they have gained more control over their schedules. On the other hand, the automation of tasks by AI can create a sense of temporal dissonance, where individuals feel disconnected from the passage of time because many aspects of their work are now handled by algorithms. This can lead to a feeling of disempowerment, as individuals lose touch with the processes that once structured their day.

Additionally, AI's ability to predict future outcomes and anticipate needs changes how we think about the future. In the past, planning for the future required active engagement and deliberation, tasks that involved the brain's recursive loops. With AI handling much of the forecasting and decision-making, individuals may become more reliant on external systems to predict future events, which could alter their perception of how much control they have over time.

Digital Dissonance: The Disconnection Between Physical and Digital Time

One of the most significant challenges of the digital age is the growing disconnection between physical and digital time. In the physical world, time is governed by natural rhythms, such as the rising and setting of the sun, the changing of seasons, and the biological need for sleep and rest. In the digital world, however, time operates on a 24/7 schedule, where information is constantly available and there is no clear distinction between day and night or work and leisure.

This disconnection between physical and digital time creates a form of temporal dissonance, where individuals feel out of sync with their natural rhythms. The brain's recursive loops, which are designed to process information in relation to biological cycles, struggle to adapt to the relentless pace of digital time. This can lead to feelings of fatigue, insomnia, and burnout, as individuals attempt to keep up with the demands of a world that never sleeps.

The Recursive Mind model suggests that this dissonance can be mitigated by developing strategies that allow the brain's recursive loops to reconnect with physical time. Practices such as digital detoxes, mindfulness meditation, and time-blocking can help individuals create boundaries between their digital and physical lives, allowing the brain to recalibrate its sense of time and restore balance. By consciously stepping away from digital devices and reconnecting with natural rhythms, individuals can reduce the cognitive strain caused by digital dissonance and improve their overall well-being.

The Role of Social Media in Shaping Time Perception

Social media platforms such as Facebook, Twitter, and Instagram have become powerful forces in shaping how we experience and perceive time. These platforms operate on a continuous loop of updates, likes, comments, and notifications, creating a sense of immediacy and urgency that accelerates our perception of time. Social media also encourages users to engage in a form of temporal comparison, where they measure their own progress and achievements against the timelines of others.

The Recursive Mind model explains how social media influences time perception by continuously feeding the brain's recursive loops with social feedback. Each time a user receives a notification or interaction, the brain loops through the social input, reinforcing the need for further engagement. This creates a cycle where individuals become increasingly attuned to the real-time feedback provided by social media, leading to a sense of time acceleration. The brain's recursive processing focuses on the instant gratification of social validation, making time feel as though it is passing quickly as users scroll through their feeds.

However, social media can also lead to feelings of temporal inadequacy. As users compare their own lives to the highlight reels of others, they may feel as though they are falling behind or not accomplishing enough in the same amount of time. This can create a form of time-related anxiety, where individuals feel pressure to constantly keep up with the achievements and milestones of their social network. The brain's recursive loops become focused on this social comparison, creating a distorted perception of time that can lead to stress and dissatisfaction.

Hypothesis: Can Digital Mindfulness Improve Time Perception in the Age of AI?

A key hypothesis that emerges from the Recursive Mind model is that adopting practices of digital mindfulness, such as setting limits on social media use, engaging in focused work without distractions, and incorporating regular breaks, could improve time perception and reduce the cognitive overload caused by the digital age. By strengthening the brain's ability to manage and filter digital inputs, individuals may be able to regain a sense of temporal balance and control.

This hypothesis could be tested by examining how individuals' time perception and well-being change before and after implementing digital mindfulness interventions. Researchers could study how reducing digital distractions, focusing on single-tasking, and creating clear boundaries between digital and physical time affect the brain's recursive processing and overall sense of time satisfaction.

Technology, AI, and Time in the Recursive Mind

In conclusion, technology, especially in the form of artificial intelligence and digital communication, has radically transformed how we experience and perceive time. Through recursive loops, the brain processes the accelerated pace of the digital world, leading to both time compression and temporal dissonance. AI's ability to automate tasks and predict outcomes creates new opportunities for time management but also introduces new challenges in maintaining a sense of agency and connection to the passage of time.

The Recursive Mind model provides a framework for understanding how the brain adapts to these technological changes, offering insights into how individuals can regain control over their time through digital mindfulness and cognitive strategies. As we move forward, we will continue to explore how the brain's recursive nature interacts with other aspects of time perception, further deepening our understanding of how time shapes human experience in the digital age and beyond.

Chapter 23: The Future of Time: Emerging Technologies and Time-Bending Possibilities

As we move further into the 21st century, emerging technologies such as virtual reality (VR), augmented reality (AR), and quantum computing are set to transform our understanding of time in ways that go beyond the accelerations already brought by digital communication and artificial intelligence. These new technologies hold the potential to bend, compress, or even expand time, creating experiences that challenge traditional concepts of how we interact with the temporal world. From simulated realities to time-bending algorithms, the future of time is set to be more dynamic, flexible, and subjective than ever before.

In this chapter, we will explore how emerging technologies could reshape our perception of time and how the brain's recursive processing will adapt to these new temporal frameworks. We will investigate the implications of virtual reality for time manipulation, how time-bending algorithms in gaming and media can alter our experience of temporal flow, and the potential for quantum computing to revolutionise time-based processes in areas such as forecasting, decision-making, and real-time problem-solving. Through the lens of the Recursive Mind model, we will examine how these technological advances may redefine the boundaries of human temporal experience.

Virtual Reality and Time Manipulation: Creating New Temporal Dimensions

Virtual reality (VR) offers one of the most profound opportunities for altering our experience of time. In a virtual environment, time can be sped up, slowed down, or even stretched in ways that are impossible in the physical world. For example, VR simulations can be designed to compress days or weeks of real-world activity into minutes, or to allow users to experience events in slow motion, heightening their awareness of each passing second. This ability to manipulate time in a virtual space opens up new possibilities for training, education, and entertainment, where users can interact with time in novel and creative ways.

The Recursive Mind model suggests that the brain's recursive loops will need to adjust to these new temporal dimensions, learning to process and adapt to environments where the normal rules of time no longer apply. In a VR simulation where time is accelerated, the brain's recursive loops may need to compress sensory and cognitive inputs to match the faster pace of the environment. In contrast, in a simulation where time is slowed, the brain may engage in deeper recursive loops, allowing for more detailed processing of each moment.

One potential application of time manipulation in VR is in training simulations. For example, medical students could practice complex surgeries in a VR environment where time is slowed down, allowing them to make careful, deliberate decisions without the pressure of real-time constraints. Similarly, athletes could use VR to simulate high-speed situations, training their brains to process information more quickly and improving their reaction times in the real world.

The brain's ability to recursively process time in these virtual environments could lead to a blurring of the boundaries between virtual and physical time. As users become more immersed in VR, their perception of time in the real world may begin to shift, with virtual experiences influencing how they understand and engage with time in their daily lives. This raises important questions about how prolonged exposure to time-manipulating environments could affect the brain's temporal calibration and whether we will need to develop new strategies for maintaining a sense of temporal balance in the future.

Time-Bending Algorithms: Reshaping Temporal Flow in Gaming and Media

Time-bending algorithms are already being used in gaming and interactive media to create experiences that defy traditional temporal boundaries. In many video games, for example, players can rewind, pause, or accelerate time, allowing them to explore different outcomes, experiment with decisions, or experience moments of heightened intensity. These algorithms enable players to manipulate the flow of time within the game world, creating a sense of agency over the passage of time.

The Recursive Mind model provides a framework for understanding how the brain processes these time-bending experiences. When players rewind time in a game, the brain's recursive loops must revisit the previous state of the game world, integrating new information from the altered timeline. This recursive processing allows players to mentally loop through different temporal possibilities, exploring alternative outcomes and learning from the results. By rewinding and replaying events, the brain engages in a form of mental time travel, where past, present, and future become flexible and interconnected.

In media, time manipulation is used to create dramatic effects, such as in films where scenes are shown in reverse or slow motion, or where multiple timelines are interwoven to create a complex narrative structure. These time-bending techniques challenge the brain's normal experience of temporal flow, forcing it to engage in more complex recursive loops to make sense of the nonlinear narrative. This requires the brain to hold multiple timelines in mind simultaneously, creating a richer, more layered experience of time.

As time-bending algorithms become more advanced, they will likely play a larger role in shaping how we interact with time in everyday technology. For example, personalised media experiences could allow users to manipulate the speed of content consumption, slowing down important moments for deeper engagement or speeding up repetitive tasks to save time. This flexibility in temporal flow could lead to a more customisable relationship with time, where individuals have greater control over how they experience different activities and events.

Quantum Computing and Time-Based Problem-Solving

One of the most exciting developments in the future of time perception is the potential impact of quantum computing. Unlike classical computers, which process information in a linear and sequential manner, quantum computers use quantum bits (qubits) that can exist in multiple states simultaneously. This allows quantum computers to process vast amounts of information in parallel, dramatically speeding up tasks such as problem-solving, data analysis, and forecasting.

The implications of quantum computing for time-based processes are profound. Tasks that once took days or weeks to complete, such as weather forecasting, molecular modelling, or financial analysis, could be completed in seconds or minutes using quantum computing. This acceleration of time-based processes will likely reshape industries such as healthcare, finance, engineering, and artificial intelligence, where real-time decision-making is critical.

The Recursive Mind model suggests that as quantum computing becomes more integrated into these fields, the brain's recursive loops will need to adapt to the increased speed and complexity of time-based processes. In areas such as forecasting and decision-making, quantum computing will allow individuals to access and process information in real-time, creating a sense of temporal expansion where they can see and anticipate future outcomes with unprecedented accuracy.

For example, in the field of medicine, quantum computing could be used to predict the progression of diseases or the effectiveness of treatments in real-time, allowing doctors to make faster, more informed decisions. Similarly, in finance, quantum algorithms could be used to predict market trends with greater precision, giving investors a clearer picture of future economic conditions. In these contexts, the brain's recursive loops will be continuously updated with new, high-speed information, allowing for more dynamic and responsive decision-making.

The Future of Time and Human Perception

As we move into a future shaped by emerging technologies, the way we experience and interact with time will continue to evolve. The brain's recursive processing of time will be challenged and transformed by new temporal frameworks, whether in the form of virtual realities, time-bending media, or quantum computing. These technologies will offer unprecedented opportunities for customising, manipulating, and accelerating time, but they will also require us to rethink our relationship with time and develop new strategies for maintaining temporal balance.

One of the most important questions we must ask is how these technologies will affect our sense of self and our connection to reality. As time becomes more flexible and subjective, will we lose touch with the natural rhythms that have long governed our lives? Or will we find new ways to integrate these technological advances into a coherent temporal experience that enhances our understanding of the world?

The Recursive Mind model offers a framework for exploring these questions, showing how the brain's ability to loop through past, present, and future experiences will continue to play a central role in shaping our perception of time, even as technologies alter the way time is structured and experienced.

Hypothesis: Can Emerging Technologies Enhance Temporal Flexibility in Human Perception?

A key hypothesis that emerges from the Recursive Mind model is that emerging technologies, such as virtual reality, time-bending algorithms, and quantum computing, could enhance the brain's ability to process time flexibly, allowing individuals to adapt to new temporal environments with greater ease. By training the brain to process nonlinear and accelerated time frames, these technologies could expand our cognitive capacity to navigate multiple temporal dimensions simultaneously.

This hypothesis could be tested by studying how individuals' time perception and cognitive flexibility change after prolonged exposure to virtual reality or time-bending media. Researchers could explore whether training in quantum-based problem-solving leads to improvements in temporal processing and future planning, providing insights into how the brain's recursive loops adapt to these new technologies.

The Future of Time and the Recursive Mind

In conclusion, the future of time is set to be shaped by emerging technologies that offer new possibilities for manipulating, bending, and accelerating time. From virtual realities that alter temporal dimensions to quantum computing that speeds up decision-making, these technologies will challenge our traditional understanding of time and create new opportunities for interacting with the temporal world. The Recursive Mind model provides a framework for understanding how the brain will adapt to these changes, offering insights into how we can navigate the future of time with greater awareness, flexibility, and control.

As we move forward, we will continue to explore how the brain's recursive nature interacts with other aspects of time perception, further deepening our understanding of how time shapes human experience in the digital age and beyond.

Chapter 24: Time and Consciousness: Synthesising Insights from the Recursive Mind Model

The relationship between time and consciousness has long been one of the most profound and challenging questions in both science and philosophy. As we have explored throughout this book, the Recursive Mind model offers a unique framework for understanding how the brain processes time and how our subjective experience of temporal flow emerges from recursive loops within the brain. In this chapter, we will synthesise the insights from previous chapters and explore how the Recursive Mind model can contribute to broader discussions about the nature of time, consciousness, and the mind's relationship with the universe.

We will investigate how time perception is central to our sense of self and identity, how recursive loops enable us to maintain temporal coherence in our mental processes, and how different states of consciousness, such as dreaming, meditation, and altered states, reveal the brain's capacity to reshape time. By examining these questions through the lens of the Recursive Mind model, we will build toward a deeper understanding of time as both a subjective and objective phenomenon, and how it shapes human experience in profound ways.

The Role of Time in Constructing Consciousness

One of the central claims of the Recursive Mind model is that time is not merely an external measure that we apply to events; rather, time is constructed by the brain's recursive loops as part of its fundamental process of creating consciousness. Our ability to experience past, present, and future depends on the brain's ability to loop through temporal information, integrating sensory input, memories, and anticipatory projections into a continuous stream of conscious awareness.

In this view, time is not simply something we perceive; it is integral to how we experience being conscious. Without the recursive processing of time, there would be no coherent sense of self, no ability to plan, no continuity in experience. Time is the framework within which consciousness unfolds, allowing the brain to construct the narrative of our lives.

The Recursive Mind model suggests that the brain's ability to process time recursively is what allows us to maintain a stable sense of self despite the constant flow of new experiences. By looping through past memories, integrating them with current experiences, and projecting possible future outcomes, the brain creates a coherent temporal framework that anchors us in the present moment while connecting us to both our past and future selves.

For example, when we reflect on a past experience, the brain does not simply retrieve a static memory; it reconstructs that memory in light of current emotions, thoughts, and contexts. This recursive process allows us to make sense of how our past experiences relate to our present identity and how they inform our future decisions. Without this continuous looping between past, present, and future, consciousness would be fragmented, and our sense of self would become disjointed.

Temporal Coherence and the Self: Building Identity Across Time

Our sense of identity is deeply tied to our perception of time. We see ourselves as existing through time, with our identity formed by a continuity of experiences that stretch from the past into the future. This temporal coherence, our ability to maintain a sense of who we are across time, depends on the brain's recursive loops, which continuously update and integrate information from different temporal modes.

The Recursive Mind model explains how this temporal coherence is maintained through the brain's ability to construct a narrative self, a mental framework that connects past experiences, present awareness, and future goals into a cohesive whole. This narrative self is not static; it is constantly being revised and updated as we encounter new experiences, emotions, and information.

For example, when we think about our life story, we are not simply recalling a list of facts about our past; we are constructing a narrative that gives those facts meaning and context. The brain's recursive loops allow us to draw connections between different moments in our lives, creating a sense of purpose and direction that extends beyond the present moment. This narrative self helps us make sense of our place in time, giving us a coherent sense of identity that persists even as our external circumstances change.

However, this process of constructing temporal coherence is not always smooth. Disruptions in time perception, such as trauma, grief, or anxiety, can cause our sense of self to become fragmented. In cases of post-traumatic stress disorder (PTSD), for example, the brain's recursive loops may become stuck, continuously replaying a traumatic event and preventing the individual from moving forward in time. This disruption in the brain's ability to process time can lead to a sense of temporal disorientation, where the individual feels trapped in the past and unable to construct a coherent narrative for the future.

The Recursive Mind model suggests that enhancing the brain's ability to process time flexibly, through interventions such as therapy, mindfulness, or cognitive training, could help individuals regain a sense of temporal coherence, allowing them to move through time with greater ease and confidence. By strengthening the brain's recursive loops, individuals may be able to overcome disruptions in time perception and rebuild a coherent sense of self.

Altered States of Consciousness and Time Perception

Altered states of consciousness, such as dreaming, meditation, or psychedelic experiences, provide fascinating insights into the brain's capacity to reshape time. In these states, the brain's normal processing of time can become altered, leading to experiences where time seems to speed up, slow down, or even disappear entirely.

The Recursive Mind model suggests that these altered states of time perception are the result of changes in the brain's recursive loops. In dreaming, for example, the brain's ability to process external sensory input is diminished, allowing the recursive loops to focus more intensely on internal experiences, such as memories, emotions, and fantasies. This shift in focus can lead to a distorted sense of time, where events in the dream seem to unfold over hours or days, even though the dream itself may last only minutes in real time.

Similarly, in meditation, individuals often report a sense of timelessness, where the brain's recursive loops become focused on the present moment, filtering out past and future distractions. This heightened state of present awareness can create a sense of temporal stillness, where the normal flow of time is suspended, and consciousness becomes anchored in the immediate experience.

Psychedelic experiences, induced by substances such as LSD or psilocybin, can also dramatically alter time perception, with users often reporting a sense of time expansion or compression. The Recursive Mind model suggests that these effects occur because the brain's recursive loops become highly plastic during these experiences, allowing for more fluid and flexible processing of time. This heightened plasticity enables individuals to experience time in ways that are fundamentally different from their normal waking state, offering new insights into the brain's capacity for temporal manipulation.

Time, Consciousness, and the Universe: Broader Philosophical Implications

At a broader level, the Recursive Mind model raises important philosophical questions about the relationship between time, consciousness, and the universe. If time is constructed by the brain's recursive loops, to what extent is our perception of time subjective, and to what extent is it tied to an objective reality? Can time exist independently of consciousness, or is it fundamentally tied to the way we experience the world?

One possibility is that time, as we experience it, is a product of consciousness, emerging from the brain's recursive processing of information. In this view, time is not an absolute, external entity but rather a mental construct that allows us to make sense of change and motion in the universe. This would suggest that our experience of time is deeply tied to the structure of the mind, and that different forms of consciousness, whether human, animal, or artificial, might experience time in vastly different ways.

Alternatively, time may be an intrinsic feature of the universe, one that the brain has evolved to process in a particular way. In this view, the brain's recursive loops are simply a tool for navigating an external, objective time, allowing us to predict and respond to the changing environment. This would imply that time exists independently of consciousness, but that the brain's ability to process time shapes the subjective experience of it.

The Recursive Mind model does not offer a definitive answer to these philosophical questions, but it provides a framework for exploring how time and consciousness are intertwined. By examining how the brain constructs time through recursive loops, we can begin to understand the profound ways in which time shapes human experience and how our perception of time reflects both the limitations and capabilities of the mind.

Hypothesis: Can Enhancing Temporal Awareness Improve Consciousness and Cognitive Function?

A key hypothesis that emerges from the Recursive Mind model is that enhancing an individual's temporal awareness, through techniques such as meditation, cognitive training, or neurofeedback, could lead to improvements in consciousness and cognitive function. By strengthening the brain's recursive loops and improving its ability to process time flexibly, individuals may experience greater clarity, focus, and self-awareness.

This hypothesis could be tested by studying how individuals' cognitive performance and self-perception change after interventions designed to enhance temporal awareness. Researchers could explore whether improving the brain's ability to process time leads to enhancements in memory, decision-making, and emotional regulation, providing insights into how time perception influences broader aspects of consciousness.

Time, Consciousness, and the Recursive Mind

In conclusion, time is central to our experience of consciousness, shaping our sense of self, identity, and our ability to navigate the world. Through recursive loops, the brain constructs a continuous flow of temporal awareness, allowing us to engage with past, present, and future experiences in a coherent and meaningful way. The Recursive Mind model offers a powerful framework for understanding how this process unfolds, revealing the complex interplay between time, memory, and consciousness.

As we continue to explore the relationship between time and consciousness, we are likely to uncover new insights into the nature of the mind, the structure of reality, and the ways in which our perception of time reflects the deep and intricate workings of the brain.

Chapter 25: Practical Applications of the Recursive Mind Model: Education, Therapy, and Technology

Having explored the deep connections between time, consciousness, and the Recursive Mind model, we now turn to the practical applications of this model in various fields. Understanding how the brain processes time through recursive loops offers valuable insights that can be applied to education, mental health therapy, technological innovation, and even personal development. By leveraging the Recursive Mind model, we can develop strategies to improve learning outcomes, mental well-being, and cognitive performance in a wide range of contexts.

In this chapter, we will examine how the Recursive Mind model can be applied to enhance educational techniques, develop new forms of therapy for addressing time-related cognitive disorders, and inform the design of technology that interfaces with the brain's recursive processing systems. These practical applications will demonstrate how the insights from the Recursive Mind model can be translated into real-world solutions that improve both individual and collective experiences of time.

Enhancing Learning and Memory in Education

One of the most promising areas for applying the Recursive Mind model is in the field of education. Learning and memory rely heavily on the brain's ability to process information recursively, looping through past knowledge, integrating it with new information, and reinforcing connections to enhance long-term retention. By understanding how recursive loops function in learning, educators can develop strategies that align with the brain's natural cognitive architecture, improving both retention and comprehension.

One key insight from the Recursive Mind model is that learning is most effective when students are given opportunities to revisit and reconstruct knowledge across different contexts and timeframes. Traditional educational models often focus on linear progression, where students move from one topic to the next without revisiting previous material. However, the Recursive Mind model suggests that learning is more effective when students are encouraged to loop back to previous concepts, revisiting and reinforcing their understanding through active recall and contextual application.

For example, a student learning about cell biology might first be introduced to the basic structure of a cell. Later, the same student might revisit this knowledge in the context of a more complex topic, such as cellular metabolism or genetic expression, allowing the brain to loop through both old and new information. This recursive approach to learning not only improves retention but also helps students develop a more integrated understanding of the material, as they see how different pieces of knowledge fit together over time.

Moreover, educators can use the spacing effect, where information is revisited at increasing intervals, to strengthen the brain's recursive processing. By spacing out learning and practice over time, rather than cramming, the brain's recursive loops have more opportunities to reinforce long-term memory pathways. This method aligns with how the brain naturally processes information in recursive cycles, leading to deeper and more durable learning outcomes.

Cognitive Therapy: Addressing Temporal Disruptions in Mental Health

Another important application of the Recursive Mind model is in the field of mental health therapy, particularly for conditions that involve disruptions in time perception or temporal coherence, such as anxiety, depression, and post-traumatic stress disorder (PTSD). These conditions often involve temporal distortions, where individuals may feel stuck in the past, unable to move forward, or disconnected from their sense of the future.

The Recursive Mind model offers a framework for understanding these disruptions as breakdowns in the brain's recursive loops. In the case of PTSD, for example, traumatic memories may be processed in a maladaptive recursive loop, where the brain continuously revisits the traumatic event without being able to integrate it into the broader narrative of the individual's life. This leads to a fragmentation of time perception, where the traumatic event feels as though it is perpetually happening in the present.

Therapeutic techniques such as cognitive-behavioural therapy (CBT), eye movement desensitization and reprocessing (EMDR), and mindfulness-based therapy can help restore the brain's ability to process time recursively. By encouraging patients to revisit traumatic memories in a controlled and adaptive way, these therapies help the brain loop through past experiences without becoming stuck, allowing for the integration of the trauma into a coherent temporal narrative. This process can help patients regain a sense of temporal continuity, reducing symptoms of PTSD and improving overall mental well-being.

Similarly, in cases of anxiety or depression, where individuals may experience a distorted or accelerated sense of time, the Recursive Mind model suggests that interventions aimed at regulating attention and enhancing mindfulness can help recalibrate the brain's recursive loops. By focusing attention on the present moment, individuals can break free from maladaptive loops that keep them focused on future anxieties or past regrets, allowing for a more balanced and flexible perception of time.

Technological Innovation: Designing Brain-Compatible Interfaces

The Recursive Mind model also has important implications for the design of technology, particularly in areas such as human-computer interaction (HCI), neurotechnology, and artificial intelligence. As we have seen throughout this book, the brain's ability to process information recursively is central to its ability to navigate complex temporal environments. By designing technology that aligns with the brain's recursive loops, we can create interfaces that are more intuitive, efficient, and compatible with human cognition.

For example, in the field of human-computer interaction, developers could design adaptive learning systems that allow users to revisit and reinforce knowledge in recursive cycles, similar to the educational strategies discussed earlier. These systems could use artificial intelligence to analyse a user's learning patterns and automatically adjust the pace and timing of information delivery, ensuring that the brain's recursive loops are optimally engaged.

In the field of neurotechnology, the Recursive Mind model could inform the development of brain-computer interfaces (BCIs) that allow individuals to interact with digital environments using their neural activity. By understanding how the brain processes time and information recursively, engineers could design BCIs that enhance the brain's natural ability to loop through temporal data, improving the speed, accuracy, and effectiveness of these interfaces.

Additionally, in the context of artificial intelligence, the Recursive Mind model offers a framework for developing AI systems that can mimic the brain's recursive processing of time. By integrating recursive loops into AI algorithms, developers could create systems that are better able to adapt to changing environments, anticipate future outcomes, and learn from past experiences. These AI systems could be applied in a wide range of fields, from autonomous vehicles to predictive healthcare, where the ability to process time recursively is critical for making accurate and informed decisions.

Personal Development: Enhancing Temporal Awareness for Greater Well-Being

Finally, the Recursive Mind model has practical applications for personal development and self-improvement, particularly in areas such as time management, productivity, and emotional regulation. By understanding how the brain processes time recursively, individuals can develop strategies to enhance their temporal awareness and improve their ability to manage time effectively.

One key strategy is to cultivate mindfulness and present-focused awareness, which helps individuals stay anchored in the present moment while maintaining a flexible relationship with the past and future. By focusing attention on the present, individuals can break free from maladaptive recursive loops that keep them stuck in negative thought patterns, allowing for greater clarity, focus, and emotional balance.

Another strategy is to practice time-blocking and interval-based work, where tasks are divided into focused time blocks followed by short breaks. This approach leverages the brain's natural preference for working in cycles, aligning with its recursive loops and improving both productivity and mental clarity.

By applying the insights from the Recursive Mind model, individuals can learn to balance their engagement with the past, present, and future, enhancing their overall sense of well-being and control over time.

Conclusion: Practical Impacts of the Recursive Mind Model

In conclusion, the Recursive Mind model offers powerful insights that can be applied to a wide range of fields, from education and mental health therapy to technological innovation and personal development. By understanding how the brain processes time recursively, we can develop strategies and tools that align with the brain's natural cognitive architecture, improving learning, well-being, and cognitive performance in diverse contexts.

As we continue to explore the practical applications of the Recursive Mind model, we are likely to uncover new ways to enhance temporal awareness, improve mental health, and design technology that interfaces more effectively with the brain's recursive loops, paving the way for a future where our relationship with time is more flexible, adaptive, and empowering.

Chapter 26: Future Research and Development: Expanding the Boundaries of the Recursive Mind Model

The Recursive Mind model opens up new frontiers for scientific research, philosophical inquiry, and technological development, offering a framework that can be applied to various domains of neuroscience, cognitive science, artificial intelligence, and even physics. As we explore the potential for future research and development, we will focus on key areas where further investigation could lead to breakthroughs in our understanding of time, consciousness, and cognitive processes.

In this chapter, we will explore how the Recursive Mind model could inform future research on the neuroscience of memory, the role of recursion in artificial intelligence, and the philosophical implications of time and consciousness. We will also consider how advances in neuroimaging technologies and brain-computer interfaces could help us better understand the recursive processes underlying human cognition and time perception. This chapter will provide a roadmap for future exploration, highlighting the areas where the Recursive Mind model holds the greatest promise for innovation and discovery.

Neuroscience of Memory: Unraveling the Recursive Loops of the Brain

One of the most promising areas for future research based on the Recursive Mind model is the neuroscience of memory. While much is already known about how memory functions in the brain, the specific mechanisms by which the brain recursively processes and updates memories remain an area ripe for further investigation. The Recursive Mind model suggests that memory is not a static process but a dynamic loop, where the brain continuously revisits and reconstructs memories, integrating them with current experiences and future projections.

Future research could focus on mapping the neural circuits involved in these recursive loops, particularly in regions such as the hippocampus, prefrontal cortex, and parietal lobes, which are known to play key roles in memory and time perception. Advanced neuroimaging technologies, such as functional magnetic resonance imaging (fMRI) and electroencephalography (EEG), could be used to observe how these brain regions interact during tasks that involve recall, planning, or future thinking. By tracking the brain's activity over time, researchers could identify the patterns of recursive processing that enable the brain to link past memories with present awareness and future goals.

Additionally, research on the plasticity of these recursive circuits could lead to new treatments for memory-related disorders such as Alzheimer's disease, amnesia, and PTSD. By understanding how the brain's recursive loops break down in these conditions, researchers could develop therapies that restore or enhance the brain's ability to process time and memory recursively, potentially reversing some of the cognitive decline associated with these disorders.

Artificial Intelligence: Integrating Recursion into Machine Learning

The Recursive Mind model also holds significant implications for the future of artificial intelligence (AI). Current AI systems are largely based on linear algorithms, where tasks are processed in a sequential manner. However, the Recursive Mind model suggests that human cognition relies heavily on recursive processing, where the brain loops through information in nonlinear and multidimensional ways. Integrating recursive loops into AI systems could lead to the development of more adaptive, flexible, and intelligent machines.

One area where recursive processing could be applied in AI is in the development of neural networks that are capable of self-reflection and self-modification. By designing algorithms that mimic the brain's ability to revisit and revise information based on new input, AI systems could become more capable of learning from experience and adapting to changing environments. This could lead to breakthroughs in areas such as natural language processing, robotics, and autonomous decision-making, where the ability to process time recursively is critical for achieving human-like intelligence.

Future research could focus on developing recursive neural networks that are capable of looping through multiple layers of information simultaneously, allowing AI systems to process complex temporal data in real-time. This could enable AI to perform tasks such as long-term planning, prediction of future outcomes, and dynamic problem-solving in ways that are more aligned with human cognition.

Additionally, recursive processing could be applied to AI systems that interface with human cognition, such as brain-computer interfaces (BCIs). By developing AI algorithms that can synchronize with the brain's recursive loops, researchers could create more intuitive and efficient interfaces for communication between humans and machines, improving the integration of AI into everyday life.

Philosophy of Time and Consciousness: Exploring Deeper Questions

The Recursive Mind model also raises important questions for the philosophy of time and consciousness, particularly regarding the relationship between subjective experience and objective reality. As we have discussed throughout this book, time is not merely an external measure but is constructed by the brain through recursive loops that link the past, present, and future into a continuous stream of consciousness. This model challenges traditional philosophical notions of linear time and invites new ways of thinking about how time and consciousness are intertwined.

Future philosophical research could explore the implications of the Recursive Mind model for questions such as whether time exists independently of consciousness or whether it is a mental construct shaped by human cognition. Additionally, the Recursive Mind model provides a framework for exploring the subjectivity of time perception, raising questions about how different forms of consciousness, whether human, animal, or artificial, might experience time differently.

One potential area of exploration is the concept of temporal flexibility, the idea that time is not experienced uniformly but can be stretched, compressed, or even suspended depending on the brain's recursive processing. This concept has profound implications for our understanding of free will, determinism, and the nature of reality itself. By exploring how the brain constructs time through recursive loops, researchers could gain new insights into the fundamental nature of time, expanding our understanding of how consciousness shapes our experience of the universe.

Neuroimaging and Brain-Computer Interfaces: Advancing Cognitive Technologies

As neuroimaging technologies continue to advance, the Recursive Mind model offers a valuable framework for interpreting brain activity related to time perception and memory. Technologies such as magnetoencephalography (MEG), optogenetics, and brain stimulation techniques could be used to further explore how the brain's recursive loops function in real time, providing insights into how these processes contribute to our sense of self and temporal awareness.

In the field of brain-computer interfaces (BCIs), the Recursive Mind model could inform the design of technologies that allow individuals to interact with machines using their brain activity. By understanding how the brain processes time recursively, engineers could develop BCIs that enhance the brain's natural ability to loop through information, improving communication, control, and interaction with digital environments.

For example, BCIs could be designed to track the brain's temporal dynamics, allowing users to interact with virtual environments in ways that are aligned with their natural time perception. These interfaces could be used in virtual reality applications, assistive technologies for individuals with disabilities, and even in cognitive enhancement devices that help individuals process information more efficiently.

Hypothesis: Can Recursive Processing Lead to a New Understanding of Time and Consciousness?

A key hypothesis that emerges from the Recursive Mind model is that recursive processing is fundamental to both time perception and consciousness, and that understanding how the brain processes time recursively could lead to new breakthroughs in neuroscience, artificial intelligence, and philosophy. This hypothesis could be tested by conducting cross-disciplinary research that integrates insights from cognitive science, neurotechnology, and philosophy, exploring how recursive loops contribute to both the subjective experience of time and the brain's ability to navigate complex temporal environments.

Future research could focus on how enhancing the brain's recursive processing abilities, through training, technology, or neurostimulation, could improve cognitive performance, enhance memory, and even expand our capacity for temporal awareness. This line of inquiry could lead to new forms of cognitive enhancement and open up new possibilities for understanding the nature of time and consciousness.

Expanding the Future of Time and Cognition

In conclusion, the Recursive Mind model offers a powerful framework for understanding how the brain processes time, memory, and consciousness. By exploring the recursive loops that shape our temporal awareness, we can gain new insights into the nature of cognition and the structure of reality. The future of research and development based on this model is vast, spanning fields as diverse as neuroscience, artificial intelligence, philosophy, and technology.

As we continue to explore these possibilities, we are likely to uncover new ways to enhance human cognition, improve time perception, and develop technologies that interface more effectively with the brain. The Recursive Mind model is not only a theoretical framework but a practical tool for shaping the future of time and consciousness in a rapidly evolving world.

Chapter 27: The Recursive Mind and the Nature of Reality: Implications for Human Experience and the Universe

As we reach the culmination of this exploration into the Recursive Mind model, it is time to reflect on the broader implications that this model holds for our understanding of reality, human experience, and the very nature of the universe. Time, as we have discussed, is not simply an external dimension that exists independently of our perception; it is constructed through the brain's recursive processes, integrating past experiences, present awareness, and future anticipations into a continuous stream of consciousness. In this final chapter, we will examine how the Recursive Mind model provides a framework for rethinking the fundamental nature of existence and how it can inform our understanding of the cosmos.

We will explore the intersections between cognition, physics, and metaphysics, considering whether time is an objective reality or a subjective construct, and how the brain's recursive processing might mirror cosmic processes in ways that offer new insights into the relationship between mind and universe. By synthesising the insights from neuroscience, philosophy, and physics, we will build toward a more holistic understanding of the role that the human brain plays in shaping not only our perception of time but also our understanding of reality itself.

The Human Brain as a Microcosm of the Universe

One of the central insights that emerges from the Recursive Mind model is the idea that the brain's recursive processing of time mirrors broader cosmic structures. Just as the brain loops through past, present, and future to create a cohesive temporal experience, the universe itself appears to operate through recursive cycles of change, renewal, and evolution. From the orbits of planets and the cycles of galaxies to the oscillations of quantum particles, the universe is filled with patterns that resemble the recursive loops seen in human cognition.

This raises a profound question: could the recursive structures of the brain and the universe be fundamentally connected? If the brain constructs time through recursive loops, and the universe also exhibits recursive patterns, it is possible that human cognition and the laws of physics are not separate realms but are interconnected through deeper, underlying principles. The brain's recursive loops may reflect the self-organising nature of the universe, where systems evolve through cycles of feedback, adaptation, and emergence.

For example, consider the way that the brain processes information in loops, constantly revisiting memories and updating them based on new experiences. This recursive process allows the brain to adapt to a changing environment, much like the universe adapts and evolves through cycles of expansion, contraction, and entropy. In this sense, the brain could be seen as a microcosm of the larger recursive patterns that govern the universe, suggesting that our experience of time is intimately tied to the fundamental structure of reality.

Time as a Construct: Objective Reality or Subjective Experience?

Throughout this book, we have explored the idea that time is not an absolute, external entity but is constructed through the brain's recursive loops. This raises a key philosophical question: is time an objective reality, existing independently of human perception, or is it a subjective experience created by the mind?

On one hand, time appears to have objective properties that are measured by clocks, calendars, and the laws of physics. In classical physics, time is treated as a linear dimension, flowing from the past to the future at a constant rate. In relativity theory, time is intertwined with space, creating the four-dimensional fabric of spacetime. These theories suggest that time exists independently of our perception and is governed by universal laws.

However, the subjective experience of time, how we perceive its passage, how we remember the past, and how we anticipate the future, is deeply influenced by the brain's recursive processes. As we have seen, the brain constructs time through loops that integrate memory, attention, and prediction, creating a fluid and dynamic experience of temporal flow. This suggests that while time may have objective properties, our experience of it is shaped by mental processes that are unique to each individual.

The Recursive Mind model therefore offers a dual perspective on time. From one viewpoint, time exists independently of consciousness, as an objective feature of the universe. From another viewpoint, time is a construct that emerges from the brain's ability to process temporal information recursively. This duality invites us to reconsider the nature of time as something that is both objective and subjective, existing both in the external world and in the internal world of the mind.

Causality and the Flow of Time: Rethinking Determinism

Another profound implication of the Recursive Mind model is its challenge to deterministic views of time and causality. Traditional views of causality suggest that the future is determined by the past, with each event leading inexorably to the next in a linear sequence. However, the Recursive Mind model suggests that time is not experienced as a strict linear progression but as a looping process where the past, present, and future are constantly being revisited and reinterpreted.

This recursive processing allows for a more dynamic and nonlinear understanding of time, where the brain's recursive loops can simulate multiple possibilities, explore different outcomes, and even alter the perception of causality based on new information. This opens the door to a more flexible view of free will, where individuals are not simply passive recipients of a predetermined future but are active agents in shaping their temporal experience.

In this view, determinism is not an absolute law but a conceptual framework that is shaped by the brain's recursive processing. While the universe may operate according to deterministic principles at a fundamental level, the brain's ability to loop through different possibilities creates a felt sense of free will, where individuals can influence the flow of time through their choices and actions. This suggests that causality is not a rigid structure but a fluid process that is constantly being reshaped by the brain's recursive loops.

The Recursive Universe: Patterns of Time in the Cosmos

If the brain constructs time recursively, could the same be true for the universe as a whole? Recent developments in cosmology and quantum mechanics suggest that the universe may operate through recursive processes that mirror the brain's recursive loops. For example, in quantum theory, particles are not fixed entities but exist in a state of superposition, where they can be in multiple states at once. This suggests that the universe operates through probabilistic loops, where multiple outcomes are possible and are only determined when observed.

Similarly, in cosmology, the idea of the cyclic universe proposes that the universe goes through infinite cycles of expansion and contraction, where each cycle gives rise to a new universe. This recursive model of the universe mirrors the brain's recursive processing, where past experiences are revisited and integrated into the present, creating a continuous cycle of renewal and adaptation.

The Recursive Mind model invites us to consider whether the recursive loops that govern time perception in the brain are part of a larger cosmic pattern, where time itself is not a straight line but a looping process that repeats across multiple scales. This recursive view of time could offer new insights into the nature of the big bang, black holes, and the multiverse, suggesting that the same principles of recursion that shape human cognition may also govern the structure of the universe.

The Recursive Mind as a Tool for Understanding Reality

In light of these implications, the Recursive Mind model serves as a powerful tool for understanding the relationship between consciousness and reality. By exploring how the brain constructs time through recursive loops, we gain new insights into the nature of memory, perception, and self-awareness. At the same time, the recursive patterns observed in the universe suggest that time may be a more dynamic and fluid concept than previously thought, shaped not only by the laws of physics but also by the mind's interaction with reality.

This holistic view of time challenges the traditional boundaries between subjectivity and objectivity, inviting us to explore new ways of thinking about causality, free will, and the universe. The Recursive Mind model offers a framework for synthesising insights from neuroscience, philosophy, and physics, allowing us to build a more integrated understanding of how time shapes both human experience and the cosmos.

Hypothesis: Can the Recursive Mind Reveal New Insights into Cosmic Time?

A key hypothesis that emerges from this exploration is that the recursive processes observed in the brain may mirror deeper cosmic principles, suggesting that the brain's ability to construct time could offer new insights into the nature of cosmic time. This hypothesis could be tested by exploring the connections between neuroscience and cosmology, examining whether the recursive loops that govern human cognition are reflected in the recursive patterns observed in the universe.

Future research could focus on cross-disciplinary studies that integrate findings from quantum mechanics, cosmology, and cognitive science, exploring whether the brain's recursive processing of time can offer new perspectives on phenomena such as time dilation, entropy, and the arrow of time. This line of inquiry could lead to new breakthroughs in both neuroscience and physics, revealing deeper connections between mind and universe.

The Recursive Mind and the Infinite Loops of Time

In conclusion, the Recursive Mind model offers a profound framework for understanding the nature of time, consciousness, and reality. By exploring how the brain constructs time through recursive loops, we gain new insights into the subjective experience of time and how it shapes our sense of self and the universe. At the same time, the recursive patterns observed in the universe suggest that time may be a more complex and multidimensional phenomenon than previously understood.

As we continue to explore the implications of the Recursive Mind model, we are likely to uncover new ways of thinking about existence, consciousness, and the cosmos, deepening our understanding of how time shapes both human experience and the universe itself. The infinite loops of time, whether in the brain or in the cosmos, invite us to explore the mysteries of reality with a sense of wonder and curiosity, knowing that the answers may lie in the recursive nature of the universe.

Chapter 28: Reflections and Future Directions: The Ongoing Journey of Understanding Time and the Recursive Mind

As we conclude this exploration of the Recursive Mind model, it is essential to reflect on the profound insights we have gained about the nature of time, consciousness, and the human mind. Throughout this book, we have delved into the intricate ways in which the brain processes time, constructing a continuous narrative of our experiences through recursive loops that link the past, present, and future. The implications of this model extend beyond neuroscience, touching on philosophy, physics, and even cosmic questions about the nature of reality itself.

In this final chapter, we will summarise the key insights from our exploration, highlighting how the Recursive Mind model provides a new framework for understanding the subjective experience of time and the role of recursion in shaping our perception of the world. We will also consider the potential for future developments in neuroscience, cognitive science, and technology, outlining how this model can continue to evolve and inform our understanding of human cognition and the universe.

Key Insights from the Recursive Mind Model

The Recursive Mind model offers several key insights into how the brain constructs time and how this process influences our experience of consciousness:

1. Time as a Construct of the Mind – Time is not an absolute, external entity but a mental construct created by the brain's recursive loops. Our perception of time is shaped by the

brain's ability to process temporal information, integrating past memories, present experiences, and future projections into a cohesive narrative. This suggests that our experience of time is deeply tied to the structure of consciousness and the brain's recursive architecture.
2. Memory and Future Thinking – The brain's recursive loops allow us to revisit and reconstruct memories, updating them based on new experiences and future expectations. This dynamic process enables the brain to maintain a continuous sense of identity and self-awareness, linking past experiences with present awareness and future goals. The Recursive Mind model highlights the importance of temporal coherence in shaping our sense of self and the ability to navigate time.
3. Cultural and Social Dimensions of Time – Time is not experienced in isolation but is deeply influenced by cultural, social, and technological factors. Different cultures and societies construct time in various ways, from linear models of progress to cyclical understandings of renewal. The Recursive Mind model shows how the brain adapts to these external time frameworks, allowing us to synchronise our actions with group dynamics and social expectations.
4. Technological Impact on Time Perception – The rapid pace of modern life, driven by technology and digital communication, has significantly altered how we perceive and experience time. The Recursive Mind model explains how the brain's recursive loops are impacted by the acceleration of time in the digital age, leading to phenomena such as time compression, cognitive overload, and temporal dissonance. Understanding these effects is critical for developing strategies to manage the demands of modern life.
5. Recursive Processing and Free Will – The brain's recursive ability to simulate multiple possibilities and loop through different outcomes suggests a more dynamic understanding of free will and causality. While deterministic principles may govern the physical universe, the brain's recursive loops enable us to experience a sense of agency and choice, allowing us to influence the flow of time through our decisions and actions.

6. The Brain as a Microcosm of the Universe – The recursive patterns observed in the brain may mirror broader cosmic processes, suggesting that the brain's ability to process time is part of a larger, self-organising system that governs both cognition and the universe. This raises profound questions about the connection between mind and cosmos, inviting further exploration of the parallels between recursive loops in the brain and recursive cycles in the universe.

The Future of Time and Consciousness Research

As we look to the future, the Recursive Mind model offers a roadmap for further research and development in several key areas:

1. Neuroscience and Temporal Processing – Future research in neuroscience should continue to explore the neural mechanisms underlying recursive processing, particularly in relation to memory, planning, and time perception. Advanced neuroimaging technologies, such as functional magnetic resonance imaging (fMRI) and magnetoencephalography (MEG), can provide new insights into how different brain regions interact during recursive processing and how disruptions in these loops contribute to conditions such as Alzheimer's disease, PTSD, and temporal disorientation.
2. Artificial Intelligence and Recursive Algorithms – The Recursive Mind model has important implications for the future of artificial intelligence, particularly in the development of recursive neural networks that can process information in more flexible and adaptive ways. By integrating recursive loops into AI systems, researchers can develop machines that are better able to handle complex temporal data, anticipate future outcomes, and learn from past experiences.
3. Therapeutic Interventions for Time-Related Disorders – The insights from the Recursive Mind model can be applied to

the development of therapeutic interventions for individuals who experience disruptions in time perception. Techniques such as mindfulness-based therapy, cognitive-behavioural therapy (CBT), and neurofeedback can be used to help individuals restore a sense of temporal coherence, improving their ability to navigate the flow of time and integrate past, present, and future experiences.
4. Technological Interfaces for Temporal Awareness – As brain-computer interfaces (BCIs) continue to develop, the Recursive Mind model can inform the design of interfaces that enhance the brain's ability to process time recursively. BCIs that track the brain's temporal dynamics could allow individuals to interact more intuitively with digital environments, improving communication, productivity, and cognitive performance.
5. Philosophical Implications of Time and Reality – The Recursive Mind model invites new philosophical inquiries into the nature of time, free will, and reality. Future philosophical research could explore the relationship between subjective time and objective reality, examining how the brain's construction of time influences our understanding of the universe and our place within it. This line of inquiry could lead to new insights into the nature of consciousness and the structure of existence.

A Final Reflection: The Recursive Nature of Time and Mind

As we conclude this journey, it is clear that the Recursive Mind model offers a powerful framework for understanding the complex relationship between time, consciousness, and the universe. By examining how the brain constructs time through recursive loops, we have gained new insights into the nature of reality and the mysteries of human cognition. The brain's ability to loop through past, present, and future experiences not only shapes our sense of self but also reflects deeper cosmic patterns that govern the flow of time in the universe.

The Recursive Mind model encourages us to think beyond the traditional boundaries of linear time and embrace a more dynamic, fluid, and multidimensional understanding of reality. It invites us to explore the infinite loops of time, consciousness, and existence, knowing that the answers to some of the most profound questions about the universe may lie in the recursive processes that shape both mind and cosmos.

As we continue to explore the implications of this model, we are likely to uncover new ways of enhancing human cognition, improving temporal awareness, and deepening our understanding of the mysteries of time. The Recursive Mind model is not only a theoretical framework but also a practical tool for navigating the complexities of the human mind and the universe, offering a roadmap for future discoveries and innovations.

Conclusion: The Infinite Possibilities of the Recursive Mind: A Journey Through Time and Consciousness

The journey we have undertaken in this book has led us through the profound intricacies of time, memory, consciousness, and the recursive nature of the human mind. The Recursive Mind model has provided a new lens through which to understand not only how the brain processes time, but also how our perception of reality, identity, and existence is deeply intertwined with the loops of cognition that form the foundation of our experience.

As we conclude this exploration, it is important to reflect on the key themes that have emerged and the broader implications of the Recursive Mind model. This model reveals that time is not merely a linear sequence of events but a dynamic construct that is actively shaped by the recursive processes of the brain. Our sense of time is created by the feedback loops between past, present, and future, enabling us to navigate the world, form memories, and anticipate the unknown.

The Significance of Recursive Processing in Human Experience

The Recursive Mind model underscores the centrality of recursive loops in nearly every aspect of human cognition. From the way we recall memories, process sensory information, or anticipate future events, the brain's ability to loop through and revise experiences is what enables us to maintain coherence in our perception of reality. These loops provide us with the capacity to adapt to changing environments, make informed decisions, and maintain a continuous sense of self despite the ever-changing flow of time.

We have seen that recursive processing not only influences our individual perception of time but also plays a critical role in how we synchronise with social rhythms, engage with cultural frameworks, and adapt to the technological accelerations of the modern world. The Recursive Mind model has demonstrated that our relationship with time is far more complex and nuanced than a simple clock-based measurement, it is deeply shaped by the brain's recursive architecture.

Time, Consciousness, and the Universe: A Holistic Perspective

Perhaps one of the most striking revelations of the Recursive Mind model is how it bridges the gap between the microcosm of human cognition and the macrocosm of the universe. By exploring how the brain processes time through recursive loops, we have discovered parallels between cognitive processes and cosmic phenomena. From the cyclical patterns observed in the universe to the oscillations of quantum particles, the recursive nature of time seems to permeate every level of existence.

This raises profound questions about the nature of reality and our place within it. Could the brain's recursive loops be a reflection of deeper cosmic principles? Is time a fundamental property of the universe, or is it a construct that emerges from the way consciousness interacts with reality? These questions remain open for further exploration, but the Recursive Mind model has laid the foundation for a holistic understanding of time that integrates insights from neuroscience, philosophy, and physics.

Practical Implications: Shaping the Future of Cognition and Technology

Beyond the theoretical implications, the Recursive Mind model offers practical applications for improving human cognition, developing new forms of therapy, and enhancing our interaction with technology. By understanding how the brain's recursive loops function, we can design educational strategies that improve learning outcomes, develop therapeutic interventions for conditions that disrupt time perception, and create more intuitive brain-computer interfaces that align with the brain's natural cognitive architecture.

The model also offers new avenues for research in artificial intelligence, suggesting that the integration of recursive loops into machine learning algorithms could lead to more adaptive and intelligent AI systems. These systems could better mimic the brain's ability to process complex information, navigate temporal environments, and anticipate future outcomes. The Recursive Mind model therefore holds great promise for shaping the future of both cognitive science and technological innovation.

The Endless Journey of Time and Mind

As we conclude, it is clear that the Recursive Mind model is not the final answer to the mysteries of time and consciousness, it is the beginning of a deeper inquiry into the infinite possibilities of the human mind. Time, as experienced through the recursive loops of cognition, remains one of the most profound and elusive aspects of our existence. It is both a subjective experience and an objective reality, shaped by the interplay between mind and universe.

The exploration of time and consciousness is an endless journey, one that will continue to evolve as new discoveries are made in neuroscience, philosophy, and technology. The Recursive Mind model has provided us with a map for navigating this journey, offering insights into how the brain constructs time and how our understanding of time shapes every aspect of our lives. Yet, like time itself, the model is dynamic, constantly evolving, revisiting, and revising its own framework as new insights emerge.

In the end, the infinite loops of time and mind invite us to remain curious, to continue questioning the nature of existence, and to embrace the complexity of the human experience. Time is not simply something we measure; it is something we live, something we shape, and something that, in turn, shapes us.

References

Neuroscience and Cognition:

1. Baars, B. J. (1997). *In the Theater of Consciousness: The Workspace of the Mind.* Oxford University Press.
2. Damasio, A. R. (1999). *The Feeling of What Happens: Body and Emotion in the Making of Consciousness.* Harcourt Brace.
3. Edelman, G. M., & Tononi, G. (2000). *A Universe of Consciousness: How Matter Becomes Imagination.* Basic Books.
4. Gazzaniga, M. S. (2008). *Human: The Science Behind What Makes Us Unique.* Harper Perennial.
5. LeDoux, J. (1996). *The Emotional Brain: The Mysterious Underpinnings of Emotional Life.* Simon & Schuster.
6. Minsky, M. (1986). *The Society of Mind.* Simon and Schuster.
7. Ramachandran, V. S., & Blakeslee, S. (1998). *Phantoms in the Brain: Probing the Mysteries of the Human Mind.* HarperCollins.
8. Schacter, D. L. (2001). *The Seven Sins of Memory: How the Mind Forgets and Remembers.* Houghton Mifflin.
9. Varela, F. J., Thompson, E., & Rosch, E. (1991). *The Embodied Mind: Cognitive Science and Human Experience.* MIT Press.

Time Perception and Consciousness:

10. Block, R. A., Hancock, P. A., & Zakay, D. (2010). How cognitive load affects duration judgments: A meta-analytic review. *Acta Psychologica, 134*(3), 330-343.
11. Eagleman, D. M. (2008). Human time perception and its illusions. *Current Opinion in Neurobiology, 18*(2), 131-136.
12. Grondin, S. (2010). Timing and time perception: A review of recent behavioral and neuroscience findings and theoretical directions. *Attention, Perception, & Psychophysics, 72*(3), 561-582.
13. Pöppel, E. (1997). A hierarchical model of temporal perception. *Trends in Cognitive Sciences, 1*(2), 56-61.
14. Wittmann, M. (2011). Moments in time. *Frontiers in Integrative Neuroscience, 5*(66), 1-9.
15. Zacks, J. M., & Tversky, B. (2001). Event structure in perception and conception. *Psychological Bulletin, 127*(1), 3-21.

Artificial Intelligence and Recursive Models:

16. Goodfellow, I., Bengio, Y., & Courville, A. (2016). *Deep Learning*. MIT Press.
17. Schmidhuber, J. (2015). Deep learning in neural networks: An overview. *Neural Networks, 61*, 85-117.
18. Mitchell, T. M. (1997). *Machine Learning*. McGraw-Hill.
19. LeCun, Y., Bengio, Y., & Hinton, G. (2015). Deep learning. *Nature, 521*(7553), 436-444.
20. Lake, B. M., Ullman, T. D., Tenenbaum, J. B., & Gershman, S. J. (2017). Building machines that learn and think like people. *Behavioral and Brain Sciences, 40*, e253.
21. Marcus, G. (2018). The algebraic mind: Integrating connectionism and cognitive science. *MIT Press*.

Philosophy of Time and Consciousness:

22. Augustine, St. (2002). *Confessions* (trans. by R. S. Pine-Coffin). Penguin Books.

23. Husserl, E. (1991). *On the Phenomenology of the Consciousness of Internal Time (1893-1917)*. Kluwer Academic Publishers.
24. James, W. (1890). *The Principles of Psychology, Volume 1*. Henry Holt and Company.
25. Kant, I. (1998). *Critique of Pure Reason*. Cambridge University Press.
26. McTaggart, J. M. E. (1908). The unreality of time. *Mind, 17*(68), 457-474.
27. Merleau-Ponty, M. (1962). *Phenomenology of Perception*. Routledge & Kegan Paul.
28. Russell, B. (1927). *The Analysis of Matter*. Kegan Paul, Trench, Trubner & Co.
29. Searle, J. R. (1992). *The Rediscovery of the Mind*. MIT Press.
30. Varela, F. (1999). The specious present: A neurophenomenology of time consciousness. In J. Petitot, F. J. Varela, B. Pachoud, & J. M. Roy (Eds.), *Naturalizing Phenomenology: Issues in Contemporary Phenomenology and Cognitive Science* (pp. 266-314). Stanford University Press.

Physics and Cosmology:

31. Barbour, J. (1999). *The End of Time: The Next Revolution in Physics*. Oxford University Press.
32. Bohm, D. (1980). *Wholeness and the Implicate Order*. Routledge.
33. Greene, B. (2004). *The Fabric of the Cosmos: Space, Time, and the Texture of Reality*. Alfred A. Knopf.
34. Hawking, S. (1988). *A Brief History of Time*. Bantam Books.
35. Penrose, R. (2005). *The Road to Reality: A Complete Guide to the Laws of the Universe*. Jonathan Cape.
36. Rovelli, C. (2017). *The Order of Time*. Riverhead Books.
37. Smolin, L. (2013). *Time Reborn: From the Crisis in Physics to the Future of the Universe*. Houghton Mifflin Harcourt.
38. Thorne, K. S. (1994). *Black Holes and Time Warps: Einstein's Outrageous Legacy*. Norton.

Mindfulness, Meditation, and Time Perception:

39. Kabat-Zinn, J. (1990). *Full Catastrophe Living: Using the Wisdom of Your Body and Mind to Face Stress, Pain, and Illness*. Dell Publishing.
40. Lutz, A., Dunne, J. D., & Davidson, R. J. (2007). Meditation and the neuroscience of consciousness: An introduction. In P. D. Zelazo, M. Moscovitch, & E. Thompson (Eds.), *The Cambridge Handbook of Consciousness* (pp. 499-551). Cambridge University Press.
41. Zeidan, F., Johnson, S. K., Diamond, B. J., David, Z., & Goolkasian, P. (2010). Mindfulness meditation improves cognition: Evidence of brief mental training. *Consciousness and Cognition, 19*(2), 597-605.
42. Wallace, B. A. (1999). The Buddhist tradition of Samatha: Methods for refining and examining consciousness. *Journal of Consciousness Studies, 6*(2-3), 175-187.

Memory and Cognitive Flexibility:

43. Baddeley, A. (2000). The episodic buffer: A new component of working memory? *Trends in Cognitive Sciences, 4*(11), 417-423.
44. Conway, M. A., & Pleydell-Pearce, C. W. (2000). The construction of autobiographical memories in the self-memory system. *Psychological Review, 107*(2), 261-288.
45. Tulving, E. (1985). Memory and consciousness. *Canadian Psychology, 26*(1), 1-12.
46. Wilson, R. A., & Keil, F. C. (1999). *The MIT Encyclopedia of the Cognitive Sciences*. MIT Press.

www.ingramcontent.com/pod-product-compliance
Lightning Source LLC
Chambersburg PA
CBHW020649220526
45464CB00001B/354